DETERRENCE AND DEFENSE IN A POST-NUCLEAR WORLD

Also by Gary L. Guertner

THE LAST FRONTIER: An Analysis of the Strategic Defense Initiative (*co-author*)

DETERRENCE AND DEFENSE IN A POST-NUCLEAR WORLD

Gary L. Guertner
Director of Studies
Strategic Studies Institute
US Army War College

St Martin's Press   New York

© Gary L. Guertner 1990

All rights reserved. For information, write:
Scholarly and Reference Division,
St. Martin's Press, Inc., 175 Fifth Avenue,
New York, N.Y. 10010

First published in the United States of America in 1990

Printed in Great Britain

Library of Congress Cataloging-in-Publication Data
Guertner, Gary L.
Deterrence and Defense in a Post-Nuclear World/Gary L. Guertner.
p. cm.
ISBN 0–312–03638–8
1. United States–Military Policy. 2. Soviet Union–Military
policy. 3. Deterrence (Strategy) 4. Nuclear arms control.
I. Title.
UA23.G95   1990
355′.033573–dc20   89–36449
                                                                                   CIP

For Sally, Brandon and Megan

Contents

List of Figures, Maps and Tables	ix
Preface and Acknowledgements	x
Glossary	xii
Introduction	1

1 Obstacles To Conventional Deterrence — 7
    Flexible Response in Europe — 9
    Convergence of Flexible Response and Forward Defense — 13
    The American Efforts To Structure Flexible Response — 14
    The Requirements For Conventional Deterrence — 18

2 Deterrence Soviet Style — 34
    Offensive Defense: Soviet Style — 35
    Stalin's Conventional Emphasis: Back To The Future? — 39
    The Legacy of Stalin — 42
    Khrushchev: Strategic Bluff or Peripheral Strategy? — 45
    The Brezhnev Period: The Origins of Flexible Options — 50
    Flexible Response Since Brezhnev — 56
    The Shrinking Soviet Nuclear Umbrella: Cover Thyself — 61
    'Reasonable Sufficiency': Revolution or Ruse? — 64
    Conclusions — 71

3 Soviet Incentives for Conventional Deterrence — 79
    Strategic Implications of the Nationalities — 85
    Implications For US Strategy — 91

4 Arms Control: Can We Make This Trip With the Russians? — 97
    The Charges — 106
    The Evidence — 110

	The Military Significance Of Non-compliance Issues	121
	Conclusions	124
5	Security in A Post-Nuclear World: The Unfinished Agenda	133
	Arms Control Strategy and Nuclear Modernization: Shaping Deterrence Stability	134
	Soviet Strategic Modernization After START	145
	Shaping Strategy and Targeting Policies For Deterrence Stability	148
	Strategic Stability and Conventional Deterrence	159
Index		175

List of Figures, Maps and Tables

Map	1.1	Forward Defense Areas and Population in the FRG	12
Figure	2.1	Soviet Strategic Evolution-ICBMs	53
Table	2.1	Weapons Destroyed Under the Terms of the INF Treaty	63
Table	2.2	Summary of Soviet Doctrinal Evolution	70
Map	3.1	Union Republics of the USSR	81
Map	3.2	Ethnic Russian Regions of the USSR	82
Table	3.1	Shifting Population Trends	83
Map	3.3	Areas of Ethnic Unrest Since Glasnost	87
Table	4.1	President's Reports to Congress on Soviet Non-compliance	107
Table	4.2	Competing Images of Soviet Non-compliance	110
Table	4.3	Military Significance of Soviet Non-compliance	122
Figure	4.1	Weapon's Cycle and Verification	128
Figure	5.1	Dilemmas of Nuclear Deterrence	135
Table	5.1	US Offensive Modernization Options	139
Table	5.2	Strategic Evolution	153
Table	5.3	Stable and Unstable Strategic Postures	157
Table	5.4	Characteristics of US and Soviet Strategic Forces	158
Map	5.1	Conventional Forces Europe (CFE) Negotiations	164
Table	5.5	Strategy After Conventional Arms Control	166

Preface and Acknowledgements

A number of trends are moving the Soviet-American strategic relationship in new directions. My optimism about these trends is neither utopian nor does it match the inflated spirit of the Reykjavik Summit which briefly had heads of state flush with enthusiasm for a world with zero nuclear weapons. I see no prospects for a nuclear-free world. It is impossible to turn back the clock. Too many technical bridges have been burned along the way; what science has given us cannot be erased from our history. There are, however, political decisions, technological innovations, and revisions of military strategy that can move us to a higher plateau on which the West and its Soviet adversary may find fewer incentives for arms competition.

Mutual deterrence between the United States and the Soviet Union is a formidable task. The verities of nuclear deterrence constantly produce policy choices that have dramatic effects on strategic stability. Examples include how we modernize offensive nuclear forces, how we proceed with strategic defense technologies, and how we combine both efforts with arms control negotiations. American efforts to extend deterrence to West European allies have made the nuclear aspect of that process more complex and volatile. Extended deterrence requires flexible military options across the conflict spectrum. In the case of Europe, conventional forces reduce the dependence on the early use of nuclear weapons that might otherwise engulf that theater in a nuclear war, which, in turn, could escalate to intercontinental attack against Soviet and American territory. The coupling of conventional and nuclear forces is intended to prevent all levels of war. The arcane strategies which flow from extended deterrence attempt to establish credible links between American infantrymen in Germany and nuclear missile silos in North Dakota and Wyoming. The nature of that linkage in the future is a central theme of this book.

Research for the book began in early 1986. Many of the issues discussed here were previously debated with colleagues and students at the US Army War College where I served as Visiting Professor of Strategy, and at the US Arms Control and Disarmament Agency where I was scholar-in-residence in 1985. Their respectful, but often healthy skepticism was a valued corrective. The RAND/UCLA Center for the study of Soviet International Behavior and the California State University, Fullerton, President's Foundation generously provided funds in support of research and writing. The book could not have been completed without their support. Conversely, there is one individual who deserves special recognition because he did everything in his power to delay the completion of the project. Mikhail Gorbachev's meteoric rise on the scene, bringing with him a revolution in military thinking, made it virtually impossible to close the manuscript. Extensive additions and rewriting were required in response to the Gorbachev phenomenon. But all revolutions eventually slow down or pause for consolidation. Whatever else the General Secretary of the Soviet Union may offer, I am satisfied that the book has captured the foundations on which future American relations with the Soviet Union and its European allies will grow.

The author wishes to thank many people who offered constructive comments on various portions of the manuscript: Robert Bresler, Dan Caldwell, Otto Chaney, Greg Govan, David Hansen, David Jablonsky, Robert Kennedy, Michael Krepon, Arthur Lykke, Jack Mendelshon, Jay Mumford, John Scott, Donald Snow, John von Trott, Mark Walsh, and Dan Whiteside.

Parts of Chapter 4 were previously published in the *Political Science Quarterly* and appear here by permission. The views in this book are those of the author and should not be ascribed to the persons or organizations whose assistance is acknowledged above.

Glossary

Arms Control – Explicit or implicit international agreements that govern the numbers, types, characteristics, deployment, and use of armed forces and armaments. (See also Arms Limitation and Disarmament.)

Arms Limitation – An agreement to restrict quantitative holdings of or qualitative improvements in specific armaments or weapons systems. (See also Arms Control.)

Assured Destruction – Ability to inflict unacceptable damage upon any single aggressor or combination of aggressors at any time during the course of a strategic nuclear exchange, even after absorbing a surprise first strike.

Containment – Measures to discourage or prevent the expansion of enemy territorial holdings and/or influence. Specifically, a US policy directed toward Communist expansion.

Controlled Counterforce War – War in which one or both sides concentrate on reducing enemy strategic retaliatory forces and take special precautions to minimize collateral casualties and damage.

Conventional Forces – Those forces capable of conducting operations using non-nuclear weapons.

Conventional Weapons – Non-nuclear weapons. Excludes all biological weapons, and generally excludes chemical weapons except for existing smoke and incendiary agents, and agents of the riot-control type.

Counterforce – A strategic concept which calls for the employment of strategic air and missile forces to destroy, or render impotent, military capabilities of an enemy force. (See also Countervalue.)

Countervalue – A strategic concept which calls for the destruction or neutralization of selected enemy population centers, industries, resources, and/or institutions which constitute the social fabric of a society. (See also Counterforce.)

Damage Limitation – Active and/or passive efforts to restrict the level and/or geographic extent of devastation during war. Includes counterforce actions of all kinds, as well as civil defense measures. Assumes a pre-emptive attack against enemy nuclear and/or Conventional Forces.

Deterrence – The prevention from action by fear of the consequences. Deterrence is a state of mind brought about by the existence of a credible threat of unacceptable counteraction; the denial of gains or the imposition of excessive costs.

Deterrence by Denial – War is averted so long as an aggressor is convinced that he cannot achieve his war objectives. This is accomplished by the ability to retaliate and/or conduct pre-emptive attacks against nuclear and conventional military forces (counterforce targeting).

Deterrence by Punishment – Similar to deterrence. Assumes a rational actor model of decision-making and mutual deterrence through assured retaliation. Assumes second-strike, countervalue targeting.

Disarmament – The reduction of armed forces and/or armaments as a result of unilateral initiatives or international agreement. (See also Arms Control and Arms Limitation.)

Doctrine – Fundamental principles by which the military forces or elements thereof guide their actions in support of national objectives. It is authoritative but requires judgment in application.

Escalation – An increase (deliberate or unpremeditated) in the scope, intensity, or geographic area of a conflict.

Essential Equivalence – A policy which stipulates a need for approximately equal capabilities, but does not demand numerical equality between the central strategic systems of the United States and the Soviet Union.

Finite Deterrence – The capability with a limited number of strategic weapons, to inflict a high level of damage (presumably unacceptable) on enemy population and industry, thus making nuclear war unthinkable.

Firebreak – A clear dividing point between conventional and nuclear weapons.

First-Strike – The first offensive move of a war. As applied to general nuclear war, it implies the ability to eliminate effective retaliation by the opposition. (See also Second-Strike.)

First-Strike Capability – The possession of sufficient nuclear weapons to eliminate the retaliatory forces of the attacked nations. It does not only mean the act of striking first.

First Use – The initial employment of specific military measures, such as nuclear weapons, during the conduct of a war. A belligerent could execute a second strike in response to aggression, yet be the first to employ nuclear weapons. (See also First-Strike.)

Flexible Response – A strategy predicted on meeting aggression at an appropriate level or place with the capability of escalating the level of conflict if required or desired.

Forward Defense – A strategy concept which calls for containing or repulsing military aggression as close to the original line of contact as possible so as to defend the entire territory of a nation or alliance.

Graduated Deterrence – A range of power that inhibits, or deals effectively with aggression at all levels of conflict. (See also Deterrence and Mutual Deterrence.)

Hard Target – A target protected to some significant degree against the blast, heat and radiation effects of nuclear explosions of particular yields. (See also Soft Target.)

Launch-on-Warning – Use of strategic weapons in a retaliatory strike upon notification or belief that an enemy has launched a strategic attack but before the enemy weapons have actually struck friendly territory.

Launch-Under-Attack – Use of strategic weapons in a retaliatory strike upon notification or belief that an enemy has launched a strategic attack and after enemy weapons have struck friendly territory.

Limited War – A war in which one or more of the belligerents voluntarily exercise restraints, for example, on the type of weapons used, the geographic limits in which the war is conducted, targets, and/or objectives.

Massive Retaliation – The act of countering aggression of any type with tremendous destructive power; particularly a crushing nuclear response to any provocation deemed serious enough to warrant military action.

Mutual Deterrence – A stable situation in which two or more countries or coalitions of countries are inhibited from attacking each other because the casualties and/or damage resulting from certain retaliation would be unacceptable. (See also Deterrence.)

Nuclear Non-proliferation – Arms control measures designed to prevent the acquisition of nuclear weapons and delivery means by nations that do not have a nuclear capability. (See also Nuclear Proliferation.)

Nuclear Parity – A condition at a given point in time when opposing forces possess nuclear offensive and defensive systems approximately equal in overall combat effectiveness.

Nuclear Proliferation – The process by which one country after another comes into possession or control of nuclear weapons. (See also Nuclear Non-proliferation.)

Glossary

Overkill – Destructive capabilities in excess of those required to destroy specified targets and/or attain specific security objectives.

Parity – A condition in which opposing forces possess capabilities of certain kinds that are approximately equal in overall effectiveness. (See also Sufficiency.)

Pre-emptive Attack – An attack initiated on the basis of incontrovertible evidence that an enemy attack is imminent.

Preventive War – A war initiated in the belief that armed conflict, while not imminent, is inevitable, and that to delay would involve greater risk. (See also Pre-emptive War.)

Second-Strike – A strategic concept which excludes pre-emptive and preventive actions before the onset of a war. After an aggressor initiates hostilities, the defender retaliates. In general nuclear war, this implies the ability to survive a surprise first-strike and respond effectively. (See also First-Strike.)

Second-Strike Capability – An essentially invulnerable nuclear force that can survive an enemy-initiated strike and still retaliate causing unacceptable damage to the enemy.

Soft Target – A target not protected or with limited protection against the blast, heat, and radiation produced by nuclear explosions or particular yields. (See also Hard Target.)

Strategic Forces – All combat forces, particularly long-range nuclear delivery systems, designed primarily for offensive strategic purposes.

Strategic Stability – A condition of the military balance that, in a crisis, offers the Soviet Union no incentive to initiate a nuclear attack. Neither is the United States under pressure to do so. By contrast, strategic instability would be a condition in which either the United States or the Soviet Union (or both) believed that victory could be achieved and defeat averted only by striking the other side pre-emptively.

Strategy – The art and science of developing and using political, economic, psychological, and military forces as necessary during peace and war, to afford the maximum support to policies, in order to increase the probabilities and favorable consequences of victory and to lessen the chances of defeat.

Sufficiency – A force planning concept which calls for adequate military force to deter attack or to prevent coercion, without costly or wasteful excess capability. (See also Parity and Overkill.)

Threshold – An intangible line separating different kinds and levels of warfare, generally the separation between nuclear and non-nuclear warfare. The greater the reluctance to use nuclear weapons, the higher the threshold.

Triad – The term used in referring to the basic structure of the US strategic deterrent force. It is comprised of land-based ICBMs, the strategic bomber force, and the Trident/Poseidon submarine fleet. These constitute the US strategic or central systems.

Introduction

This book is about American and Soviet strategies for deterrence and war. Its title is boldly optimistic. Its focus, however, is not on the arms control end game–a world entirely free of nuclear weapons. A post-nuclear world will not be a nuclear 'free' world. Nuclear weapons in reduced numbers are certain to remain in the arsenals of the superpowers, but they will play a declining role in military strategy and war plans.

The search for security in the nuclear age has opened three paths that must be traveled simultaneously: (1) Deterrence; (2) Arms control; and (3) Strategic defense. Each path has its champions and critics. In fact, neither superpower can afford to turn away from any one of these paths, nor race unimpeded down any single one.

The arms control path is widening more rapidly than anyone could have imagined when Ronald Reagan became president: an INF Treaty has been signed; a START agreement is possible during the Bush administration; and conventional arms control is high on the agenda. Arms control proponents have reason for optimism. Yet the transition through which we are living may be the most dangerous and unstable period in the nuclear age. Extricating ourselves from an era of strategic instability will require not only arms control agreements, but also breakthroughs in military technology, both offensive and defensive.

Technology may indeed succeed in making ballistic missiles 'impotent and obsolete', but not because that has been a goal of the Reagan Strategic Defense Initiative. The search for non-nuclear defenses under the SDI will inevitably, as the Soviets have always feared, breed a new generation of offensive weapons. Weapons of mass destruction will give way to complex systems capable of great destruction, but with precise discrimination between military and non-military targets. Military strategy in space and for land warfare will be developed around emerging technologies that turn 'smart' weapons into 'brilliant' weapons. No trend is more predictable given the imperatives of military strategy and the inexorable pace of technology.

These forces are driven by a simple truth. Powerful nations must eventually base their military strategies on weapons that can be used. This overwhelmingly simple strategic logic will push us into the post-nuclear era. The process has been under way for at least two decades in the form of evolving Soviet and American strategies that have sought usable military options in the nuclear age. 'Flexibility', 'flexible response', 'limited nuclear options', 'damage limitation', 'pre-emption', and 'denial', are strategic concepts that have raised a common question–Can deterrence and the threat of war be made credible with nuclear weapons? The strategic evolution described here suggests that the question cannot be answered affirmatively. Trends toward non-nuclear options will accelerate dramatically in the next decade as a result of technology, arms control, and the impossibility of marrying nuclear deterrence and nuclear war fighting in one coherent doctrine. US/NATO nuclear policies are the most dramatic example of this dilemma.

How we deter nuclear war between the United States and the USSR is intimately related to our strategy for deterring war in Europe. Extending American military might to the defense of its NATO allies has been the driving force in American defense policies. Our military strategy, our force structure, and the anxieties that surfaced on both sides of the Atlantic during the INF Treaty debate reflect this dominant relationship.

Military strategy is driven by the mechanics that link American and European security. These linkages have changed dramatically since NATO's creation in 1949 and, over time, nuclear weapons have played a diminishing role. Deterrence in the Truman and Eisenhower Administrations was simple and lethal. If the Soviet army invaded Europe, America would retaliate with nuclear weapons against Soviet urban/industrial targets. Deterrence by threat of punishment was insufficient to those who doubted that strategic nuclear attacks against the Soviet homeland could stop an avenging Soviet army before it occupied Europe.

Once the Soviets deployed their own arsenal of theater and strategic nuclear weapons, the credibility of one-sided nuclear deterrence was lost forever. By the early 1970s, the Soviets had achieved strategic nuclear parity and deterrence became unambiguously mutual. If it failed, assured destruction also became mutual. These developments created a need for greater flexibility in American military strategy. Deterrence of war would be credible only if military strategy extricated itself from the trap of mutual assured destruction. If deterrence failed, it was hoped that wars would be fought at their lowest level of violence. The search for flexibility took place at two distinct levels–in European and in US plans for conducting a strategic nuclear war with the Soviet Union.

In Europe, raising the nuclear threshold through conventional war fighting capabilities was the major objective of flexible response. But a divisive question remained unanswered. Should nuclear deterrence be subordinate to the primary task of conventional defense? This book argues for the primacy of conventional defense in an era of Soviet-American nuclear parity. Extending American military power to the defense of Europe can and should be done, but without linking it to the strategic nuclear umbrella. Strategic coupling, the threat to escalate war in Europe to intercontinental nuclear attacks against Soviet territory, is no longer–if it ever was–a prudent strategy for Americans or a reasonable expectation for Europeans.[1]

The second form of flexible response discussed in these pages is the American effort since the 1970s to develop limited nuclear war fighting options. Massive nuclear exchanges, it was argued, were no longer credible. Deterrence, therefore, could be credible only if more limited options were available. If deterrence failed, it need not fail catastrophically, leading to what the late Herman Kahn described as 'spasm warfare'. Limited or small-scale nuclear attacks against military targets, intra-war bargaining, threats of escalation if aggression continued, and conflict termination short of full-scale nuclear war became major components of strategic flexible response or limited nuclear options.

To avoid confusion, flexible response is used here to describe NATO strategy in the European theater. Flexible responses in the conduct of strategic nuclear war between the United States and the Soviet Union are referred to as limited nuclear war. These two levels of flexible response–theater and strategic–are intimately connected. A strategic nuclear exchange as a bolt-out-of-the-blue is unlikely. Nuclear war is more likely to grow out of conflict in the European theater. Hence, the link between NATO strategy of flexible response and US strategy for deterring a strategic nuclear attack on the continental United States. The importance of this link was dramatically underscored during the debate to deploy (and remove) American long-range theater nuclear weapons (LRINF) to Europe. These missiles, it was argued, filled a gap in NATO strategy. If conventional forces or battlefield nuclear weapons (SNF) failed to deter or defeat the Warsaw Pact, nuclear missiles with a range that threatened Soviet territory could. American long-range nuclear weapons in Europe relieved European anxieties over the credibility of an American nuclear umbrella that was dependent on strategic nuclear forces deployed in the United States or at sea. Attacking the Soviet Union with intercontinental weapons risked Soviet retaliation. Europeans asked if an American president would risk New York for Bonn? For Europeans, long-range American nuclear weapons in Europe strengthened deterrence by removing this nagging question. Deterrence was further strengthened in European eyes by Soviet threats to retaliate directly against US territory if the USSR was hit by American nuclear weapons based in Europe. Whether the Soviets would retaliate or not, Europeans believed that LRINF strengthened deterrence, and relieved pressure for more investment in costly conventional forces.

American interests are served only if theater nuclear weapons provide NATO's nuclear umbrella without the risk of direct attacks against the United States. The analysis of Soviet military strategy described here–and Soviet reactions to intermediate-range missile deployments–show clearly that the Soviets will not tolerate a one-sided nuclear war. Neither are they likely to conduct escalation (retaliation in their view) at the slow tempo prescribed in the US strategy of limited nuclear warfare. Long-range nuclear weapons in Europe, therefore, created an escalation trap if deterrence failed. The INF Treaty extricates the United States from this trap and removes a major catalyst to intercontinental nuclear war. It also separates the strategies for defending Europe from those required to deter a nuclear attack against the United States. Keeping these objectives separate is in the American interest. A successful strategy for doing so will depend on the combined effects of conventional force modernization and arms control. These are the issues developed in *Deterrence and Defense in A 'Post-Nuclear' World*.

Although the book's early emphasis is on the European theater, its scope is broader, encompassing the strategic nuclear relationship with the Soviet Union. Chapter 1 describes the efforts within the NATO alliance to build a credible deterrent for Europe. Coalition deterrence rests on three strategic concepts–forward defense, flexible response, and strategic coupling. In practice, the uneasy compromises between American and European interests have produced a volatile strategy. Forward defense or fighting as close to the inter-German border as possible limits flexible response through its requirements for early use of nuclear weapons. Early use of nuclear weapons, Soviet escalation in kind, and the coupling of American strategic nuclear forces to NATO's defense (the umbrella) combine to produce an escalation trap if deterrence fails.

Chapter 2 develops this theme further by examining the evolution of Soviet military doctrine and strategy. Soviet military strategy has evolved, like its American counterpart, in response to changes in technology, improvements in force structure, and above all, with the achievement of nuclear parity with the United States in the 1970s. Flexible response Soviet style is visible in a strategy and force structure that shows a clear preference for conventional war in Europe. Unlike their NATO counterparts, however, Soviet strategists are less concerned with escalation than with the level at which war can be decisive. From Brezhnev to the present, Soviet military doctrine has tasked its military forces to act as both sword and shield. Strategic nuclear forces are the shield that deters nuclear attack against Soviet territory; conventional forces are the sword, available to fight and win a decisive war in Europe before NATO's nuclear weapons become a threat along a nuclear path leading to Soviet territory.

Chapter 3 highlights Soviet strategic vulnerability as a multinational state with complex and unique requirements for homeland defense. Soviet vulnerabilities provide both a requirement and an incentive for stability, arms control, and non-nuclear options in their military strategy.

Chapter 4 assesses the reliability of the Soviet Union as an arms control partner by looking at its compliance record with SALT I, SALT II, and the ABM Treaty. Three interpretations of Soviet behavior are assessed: bureaucratic politics among the Party-State-Military infrastructure that may have resulted in imperfect supervision of treaty obligations; premeditated cheating strategies to gain military advantage; and 'normal' contractual disagreements over the terms and meaning of treaties similar to the kinds that employ thousands of lawyers in the domestic economy. Alternative interpretations of Soviet behavior, marginal military gains from disputed activities, and the intrusive on-site inspection procedures accepted by the Soviets in the INF and START Treaties weaken the case against expanding the existing arms control regime.

The final chapter explores specific arms control problems required to achieve stability between Soviet and American strategic nuclear forces and in the European theater. The most contentious issues will be the offensive-defensive relationships confronting negotiators in the arenas of START-SDI and in conventional arms control efforts to structure a 'non-offensive' defense. Negotiations, even if they fall short of expectations, will contribute to the momentum toward non-nuclear options in military strategy and future weapons development.

Those readers looking for a detailed description of the future will be disappointed. The book lives in the present. The non-nuclear future will emerge slowly, when and if we solve the problems described here. Ironically, success is unlikely to reduce the frequency of conflict and war. They too will remain, and the use of force may become even more frequent if military power breaks free of its nuclear bonds.

The visible trend toward high-tech conventional forces will accelerate. Deterrence and defense in the future will be dominated by states who capture the technological highground. The Strategic Defense Initiative may serve the nation's interests in a way that was never anticipated. Its most lasting contributions may come through technologies that can be applied to non-nuclear offense. The law of unanticipated consequences may open the door wider to the post-nuclear world–for better or worse.

NOTE

1. The author was surprised to find how far this thinking has evolved in official US defense policies. See, for example, Secretary of Defense Frank C. Carlucci's *Annual Report to Congress, Fiscal Year 1989*. He states, 'As the Soviet capability to inflict massive damage on the United States' homeland has grown, our early resort to nuclear weapons in a conflict has become less desirable. While the United States and its allies reserve this option, the alliance cannot rely on it too heavily'. P. 55.

1 Obstacles to Conventional Deterrence

> When they get a bomb to neutralize our bomb, we better have an army to neutralize their army . . .
>
> General Omar Bradley

More than two decades have passed since flexible response and non-nuclear war fighting options were formally accepted as the core of NATO strategy. Raising the nuclear threshold through credible conventional defenses for Europe has been central to that strategy. The Soviet buildup in strategic nuclear weapons in the early 1970s, followed by equally dramatic theater nuclear force modernization programs, resulted in an overly narrow, nuclear focus on NATO strategy in Europe. Soviet acceptance of the American proposal to eliminate intermediate-range nuclear missiles and progress in conventional arms control have refocused Western attention on NATO's conventional forces. NATO strategy will undergo dramatic changes if conventional forces are reduced in accordance with the ambitious proposals coming out of both Washington and Moscow. As the difficult details of a conventional arms treaty emerge, it is important to remember how and why NATO's 1990 force levels and military strategy evolved in the first place.

Both superpowers in their own way have developed flexible strategies in the European theater. Ironically, at the outset of conventional arms negotiations it was the Soviet Union that was most prepared through its military strategy and force modernization programs to execute flexible response, and to fight a decisive conventional war in Europe.

Applying Western concepts of flexible response to Soviet military doctrine and strategy requires careful delineation of how the Soviets view war. American strategists describe flexible response as a political-military decision to escalate hostilities from one level of military effort to another of greater magnitude when confronted with the possibility of defeat or escalation by enemy forces. This includes escalation from conventional to nuclear weapons (vertical escalation), and escalation of hostilities from the European theater to direct attacks against Soviet territory. The essence of flexible response, in theory, is the initial defense of Europe with conventional forces. In practice, political requirements within the alliance have resulted in a more ambiguous strategy. Flexible response has not resulted in precise timetables for escalation from conventional to nuclear war. NATO's exact response to aggression confronts Soviet planners with three possible responses if deterrence fails: (1) 'Direct defense' by conventional forces; (2) the threat of escalation to nuclear weapons; and (3) the threat of retaliation against Soviet territory. Confronted with a credible NATO force structure, Soviet leaders cannot be certain which level of response their actions, even if limited to conventional forces, might trigger.[1]

The Soviets view escalation, and therefore flexibility, in terms theoretically different from Western concepts which more explicitly identify thresholds or fire breaks separating one level of war from another. War is a deadly serious business. It should not be surprising that the Soviet's critical threshold is war itself once they are directly involved.[2] Once the decision has been made (forced on them according to Soviet military writings), *decisive* action to achieve a swift military defeat of enemy armed forces is more critical in Soviet military strategy than the specific levels of intensity or thresholds between conventional and nuclear weapons. Bold action and initiative in the early stages of war are more important than Western concepts of escalation control, intra-war bargaining, and conflict termination.

This does not mean that Soviet leaders do not recognize the dangers or destructiveness of nuclear weapons, or that they are not prepared to fight a conventional war in Europe. Indeed, Soviet military writings and weapons deployment patterns since the 1970s indicate that they now see a war in Europe without the use of nuclear weapons as a distinct and *desirable* possibility. This shift from previous Soviet insistence that war in Europe would inevitably be nuclear has little to do with Western-style concepts of *flexibility* or escalation. Soviet strategy and force posture make it reasonably clear that a conventional phase of war in Europe may be possible only to the extent that it could be decisive in achieving a military victory over NATO forces. There is the distinct probability that nuclear weapons will be used, and Soviet forces are prepared to use them. Their employment, however, like Soviet conventional forces, is guided by military considerations of decisive impact on the war in progress rather than by Western-style political calculations of escalating to a new level of military intensity where defeat of the enemy *may* still be possible short of a full commitment of available force.

The NATO *minimalist* (flexibility) and the Soviet *maximalist* (decisiveness) approaches to the initial stage of war have grown out of different views of what deterrence may require, how war should be fought when deterrence fails and, in the case of NATO, inevitable political compromises in structuring coalition deterrence and warfare.

FLEXIBLE RESPONSE IN NATO: THE EVOLUTION OF COALITION DETERRENCE

Broadly speaking, the objectives of flexible response are the same on both sides of the Atlantic–the deterrence of war in Europe. Narrower, but persistent political pressures have, however, resulted in considerable differences between American and European concepts of flexibility. This is most notably true in the case of West Germany.

On the American side, flexible response grew out of a search for credible deterrence in Europe that raised the nuclear threshold. From the beginning, the West German interest in flexible response has been subordinated to the political requirements of forward defense and the strengthening of deterrence through declaratory policies which are unyielding in their defense of West German territory.

Flexible response was developed during the McNamara era as a means of reducing the reliance on nuclear weapons by fielding credible levels of conventional forces that could, depending on their success on the battlefield against invading Warsaw Pact forces, be supported by firepower ranging from conventional weapons to battlefield, theater, and strategic nuclear weapons. Each level of response represented carefully planned and controlled decisions in the escalatory process.

The German view of flexible response has never embraced the American concept of flexibility as a continuum of military responses from conventional to nuclear. Instead, the Germans believe deterrence is strengthened if the Soviets are confronted with ambiguity. Flexibility, therefore, means Soviet inability to calculate risks provided that NATO strategy leaves open the question of when and how it might respond. The first Soviet tank to cross the inter-German border may face NATO combined arms, including an early nuclear response. For most Germans, flexible response is a choice of firepower that could be applied against Warsaw Pact forces at any time, and not an escalatory process to be dictated by progress on the battlefield.[3]

These two versions of flexible response grew out of distinct American and German strategic interests. They also reflect considerable debate over which NATO *declaratory* policies strengthen deterrence most (of great interest to Germans) and which *employment* or war fighting strategies would be most effective in defeating an attack if deterrence fails (of great interest to Americans).

Whatever relative merits these two versions of flexible response may have, NATO strategy has, in fact, evolved closer to the German view. This evolution was more directly the result of political compromise than of direct military planning. The most important of these constraints has been the reluctance of Europeans to spend more on conventional forces or to move away from their reliance on the threat of early nuclear response as a deterrent.

Equally important, but more frequently neglected by analysts as a force that has shaped NATO's strategy, is the German insistence on 'forward defense'. Forward defense envisages almost all of NATO's ground forces in West Germany lining up in a narrow band of territory extending roughly 20–50 kilometers behind the inter-German border. A typical NATO Corps will deploy a covering force at the border to harass and delay the lead elements of an advancing force, giving time for the main body of the defense to deploy and prepare positions 20 to 30 kilometers to the rear. A Warsaw Pact penetration beyond 50 kilometers of the border would be regarded as a breakthrough of NATO's main line of defense.[4]

Germans are understandably reluctant to give up territory or see a NATO-Warsaw Pact confrontation on their territory where the American distinctions among battlefield, theater, and strategic nuclear weapons would be lost on West Germans residing on or near the battlefield. These fears that motivate the German preference for forward defense were described in detail by the West German Minister of Defense in 1976:

> In view of their high degree of industrialization and the density of their population and the consequent vulnerability of all their state machinery, the NATO countries of Western Europe are hardly in a

position to incur territorial losses without jeopardizing their existence. This is particularly true of the Federal Republic of Germany, situated along the dividing line between NATO and the Warsaw Pact in the center of the field of political tension. Here, the great concentration of population and economic potential is marked by 24 areas of density, in which 45 percent of the population and even 55 percent of those gainfully employed live and work–in fact in an area comprising only 7 percent of federal territory. These areas of density are expanding still farther and accrete along the main traffic arteries to form urbanized corridors. Such corridors and broad settlement areas limit extensive military movements. This applies to both the attacker and the defender. Moreover, the structure of the settlements makes the control of comparatively large formations difficult.[5]

Clearly, West Germans did not create their 'urban corridors' as military obstacles to blunt a Soviet *blitzkrieg*. They are considered vital centers that must be kept free of destruction. If NATO's forward defenses were to fail and allied forces fell back deliberately to gain time and reinforcements, fighting would inevitably approach Germany's urbanized corridors. Since millions of Germans live and work in these areas, fighting could quickly engulf them. Space is too precious to trade for time (see Map 1.1).

As compelling as forward defense is for German domestic politics, it does not necessarily serve the American (and arguably not the German interest in deterrence) strategic interest of keeping a war in Europe conventional for as long as possible. Forward defense lowers the nuclear threshold and risks early escalation to a Soviet-American nuclear exchange. It has been the driving force in pulling NATO strategy from the American to the German concept of flexible response. Forward defense, in its most rigid application (early reliance on the first use of nuclear weapons) robs flexible response of its credibility making it appear more like 'Russian Roulette' than a strategy that incorporates escalation control and conflict termination short of a general nuclear exchange. A robust flexible response capability requires equal flexibility in forward defense. The battlefield in Central Europe must have depth, and conventional forces must have reserves, mobility, fall-back positions, and lateral striking power if they are to play a credible role in NATO strategy. As one US commander recently observed, 'Forward defense means that you start the forces well forward. It doesn't mean you have to die there'.[6]

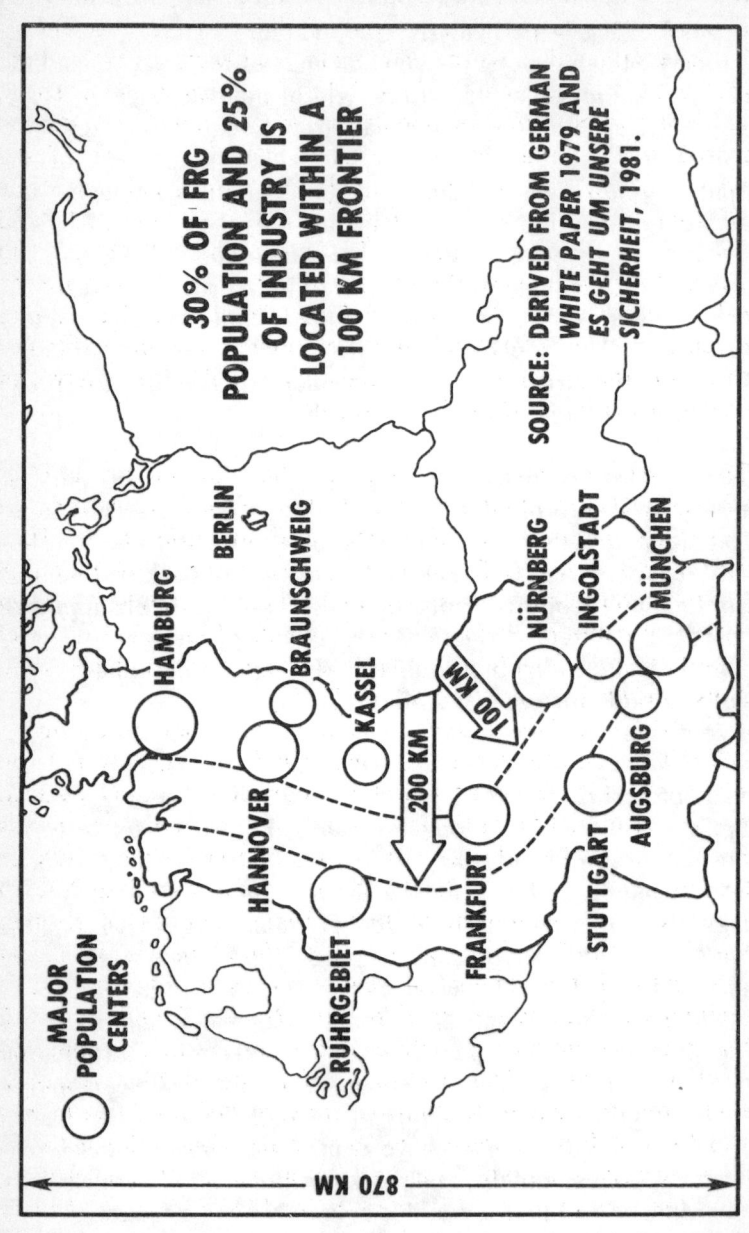

MAP 1.1 *Forward defense areas and population in the FRG*

Where the battlefield is will determine *how* war in Europe will be fought.

CONVERGENCE OF FLEXIBLE RESPONSE AND FORWARD DEFENSE[7]

American occupation forces in Germany were stabilized at a low point of 135 000 men in July 1947. While American military planners were reorienting their thinking from the original occupation mission to defense against a Soviet attack, wide-ranging discussions emerged over where that defense might actually take place. Arguments were heard from a number of sources, contending that the line ought to be drawn variously at the Vistula Oder-Neisse, Elbe, Rhine, Brittany Peninsula or Pyrenees Mountains. A majority of Europeans on the continent favored halting aggression as far forward as possible. French Premier Henri Quelle graphically pleaded for a defense at the Elbe.

> We know that once Western Europe was occupied, America would again come to our aid and eventually we again would be liberated. But the process would be terrible. The next time you probably would be liberating a corpse.[8]

Once the North Atlantic Treaty was signed on 4 April 1949, the United States committed itself publicly to the concept of forward defense. Declaratory policy turned out to be substantially different, however, than actual military planning which had to face up to the realities of the military balance that favored the Soviet army.

British and American planners favored meeting the forward defense pledge through screening and delaying forces between the Elbe and the Rhine, with the bulk of NATO forces employed in defensive positions along the Rhine River. This force was to hold at the Rhine until American and European reserves could be mobilized and employed in a counter offensive. The French favored a stand closer to the inter-German border. The difference in definitions of forward defense amounted to as much as 150 kilometers of West German territory. These differences were also reflected at the North Atlantic Council in its September 1950 meeting. This meeting extended NATO protection to Germany and formally adopted the doctrine of forward defense, but without specifying where 'forward' was to be on the ground.[9]

The allies could agree on broad declaratory principles, but remained silent on the operational specifics required to implement forward defense. Military planners were allowed wide latitude for interpretation. Americans spoke in terms of mobile rather than linear defenses, and General Bradley, Chairman of the Joint Chiefs, made references to the great depth of the NATO battlefield.[10] For the Germans, the September 1950 decision marked the end of a Rhine defense and the beginning of allied abilities to mount a forward defense beginning at their border.[11] This meeting may well have been the origin of German-American differences over forward defense, differences that were glossed over in declaratory policies, but which became more divisive in operational planning. This was specially true after the introduction of nuclear weapons to NATO in 1953. The closer to the border NATO defenses were mounted, the more important early use of nuclear weapons became, and the less flexible NATO's response could be.

By late 1952, planners at SHAPE had failed to convince their European critics that the concept of a mobile defense between the Rhine and the inter-German border was the most forward defense possible within the limitations of the conventional balance of forces. Under the pressure and competing demands of a political strategy which proclaimed forward defense and military strategy that required credible plans and force levels, NATO planners began to look seriously at battlefield nuclear weapons as the primary guarantor of defense at the border. These plans evolved by 1954–56 to a posture of early nuclear response to a Soviet attack in the Central European theater. This first use strategy resulted in the role of conventional military forces being modified from one of maneuver to trip-wire. That is, manpower requirements were limited to those levels needed to demonstrate American resolve, force the attacker to mass his forces, and to deliver nuclear firepower against Soviet invaders if deterrence failed.[12]

THE AMERICAN EFFORTS TO STRUCTURE FLEXIBLE RESPONSE

The strategic doctrine of the 1950s which emphasized deterrence through early and massive use of nuclear weapons (massive retaliation) was openly challenged by the Kennedy administration. In the Pentagon, McNamara established his leadership through a memorandum consisting of 96 questions designed to prod the Pentagon bureaucracy into re-examining old assumptions and policies. One of those called for the development of a flexible response doctrine for strategic nuclear forces. At the strategic level, flexible response meant the beginning of a long transition away from massive, spasm attacks against Soviet cities and military targets, toward more measured responses against military targets. These efforts, which continued into the Reagan era, were designed to give strategic deterrence greater credibility and to reduce the destruction of nuclear war if deterrence failed. The search for a credible nuclear strategy gradually incorporated the concepts of limited nuclear war, escalation control, intra-war deterrence, and conflict termination.[13] Changes in American strategic doctrine were accompanied by a re-examination of NATO strategy. Raising the nuclear threshold in Europe meant that conventional forces would have to play a new and more critical role in both deterrence and military planning.

The arguments for relying on tactical nuclear weapons as substitutes for manpower (some 7000 tactical nuclear weapons had been deployed by 1968) were unconvincing to the analysts brought into the Pentagon by McNamara. Their now familiar critique emphasized:

- Every Pentagon war game that had studied the problem of tactical nuclear warfare had concluded that far from rectifying conventional manpower imbalances, tactical nuclear war could, through higher rates of attrition, exacerbate those imbalances without reducing the importance of conventional manpower to the final outcome of a war in Europe.
- Because the central front was heavily congested with a growing urban/surburban population, warfare begun with the intention of limiting tactical nuclear warfare to discrete military objectives would quickly devastate civilian societies, thereby increasing the risk of escalation on the Continent, which in turn would risk escalation to central strategic nuclear war.
- The vulnerability of forward-based nuclear weapons would create instabilities during crises in the form of pressure to use weapons before they were destroyed or fell into enemy hands.[14]

From these theoretical critiques emerged the arguments for establishing 'firebreaks' (that is, flexible response) between nuclear and conventional operations in NATO strategy. Proponents of NATO's tactical nuclear posture argued that it strengthened deterrence by introducing costs that outweighed potential gains, and thereby helped dissuade Warsaw Pact leaders from initiating an attack. And what was the alternative? The Soviet Union had 175 divisions on the ground facing some 25 undermanned NATO divisions. NATO would require more than 50 divisions on the central front to meet this overwhelming Soviet threat. This kind of buildup would, it was estimated, require at least a 20 per cent increase in spending for conventional forces that might result in nothing better than a replay of World War II.[15]

It was conventional wisdom which imputed a superiority to the Warsaw Pact so great as to be unchallengeable by the West. If NATO strategy was to be pulled away from its reliance on nuclear weapons toward a credible conventional posture, it would be necessary to find flaws in the existing assessment of the military balance. McNamara did this with astonishing results.

Analysts at the Defense Department's Office of Systems Analysis and International Security Affairs (ISA) staff headed by Paul Nitze set out to re-examine the conventional wisdom. Comparisons of aggregate population, GNP, levels of technology, manpower advantages, levels of military spending, comparative combat division size, readiness, and combat effectiveness all pointed in one direction:

> The magnitude of Warsaw Pact conventional superiority had been grossly exaggerated. In fact, eliminating paper divisions, using cost and firepower indexes, counts of combat personnel in available divisions, and number of artillery pieces, trucks, tanks, and the like, we ended up with the same conclusion: NATO and the Warsaw Pact had approximate equality on the ground. Where four years earlier it had appeared that a conventional option was impossible, it now began to appear that perhaps NATO could have had one all along.[16]

McNamara used these Pentagon studies as justification for increasing US conventional strength. The Army was increased from 11 to 16 active divisions (18 today); tactical fighter wings increased from 16 to 21; logistical networks were improved; airlift capacity was increased; and the first prepositioned sets of division equipment were sent to Europe to aid in rapid reinforcement. These measures were taken in part because of the importance the administration placed on enhancing US conventional capabilities, and in part because revised estimates of the military balance showed that the efforts to raise the nuclear threshold by building a credible conventional capability were not futile. Selling the new strategy to Europeans, however, was far more difficult.

McNamara outlined the emerging US strategy of flexible response on the strategic nuclear level as an effort to make the nuclear deterrent more credible. If deterrence failed, a shift toward military targets could mitigate the consequences of nuclear war, and possibly result in war termination at lower levels of conflict and destruction. The important linkage between the new American strategy and European security was stressed in unprecedented detail. McNamara made it clear that NATO could rely on American strategic weapons to cover major Soviet nuclear forces threatening Europe. US operational planning (the SIOP or Single Integrated Operation Plan) revealed by the Secretary of Defense showed the degree to which the American nuclear guarantee had actually been written into the operational plans for the alliance. Military targets in Eastern Europe as well as in the Soviet Union were, as they are now, well covered by American long-range, strategic forces. Strategic coupling (that is, the American nuclear umbrella) meant new vitality for deterrence in Europe, because nuclear war would have no effective geographic boundary which could restrict nuclear war to Europe. Strategic coupling was an explicit threat that Soviet aggression could quickly escalate to a Soviet-American strategic nuclear exchange. At the same time, the danger of escalation gave a new urgency to the role of conventional forces as a 'firebreak' separating nuclear from conventional warfare. It is important to emphasize here that the price McNamara was exacting from Europeans for the American nuclear umbrella was to be a credible conventional defense with major European participation. Strategic coupling through nuclear weapons alone was risky. By the end of the decade it was not only risky, but suicidal in the face of growing Soviet strategic parity. American security required conventional forces to reduce those risks. This quid-pro-quo quickly became lost in political compromises hidden in the ambiguities separating declaratory policies from actual military strategy and war plans. Twenty years later, the INF Treaty has forced a painful re-examination of these same issues.

The elements of flexible response were written into major NATO planning documents, but a more limited and ambiguous doctrine than McNamara envisioned resulted. The version finally adopted in 1967 made the threat of deliberate escalation explicit. The threat of escalation from one level of warfare (conventional) to another (nuclear) is a crucial element in NATO's declaratory policies. Actual employment policies or war fighting plans, however, are far more ambiguous because of the critical temporal elements of strategy. How soon and under what circumstances will escalation take place? How early should NATO consider introducing nuclear weapons into conventional operations? The answers to these questions raise politically divisive questions. What is clear, however, is that NATO strategy, according to former Supreme Allied Commander in Europe, General Bernard Rogers, would force escalation 'fairly quickly to the first use of nuclear weapons to halt a large scale attack by Soviet conventional forces in Europe'.[17] Flexible response may not be very flexible at all, because NATO relies heavily on its nuclear capabilities. This reliance is primarily the result of continued European opposition to spending more on conventional forces and on the German mandate for forward defense. The end result is a strategy that more closely parallels Soviet declaratory policies of the early 1960s which assumed that any conflict involving the superpowers would inevitably and quickly escalate to nuclear war.

THE REQUIREMENTS FOR CONVENTIONAL DETERRENCE

Few believe that war in Central Europe is likely. The risks of miscalculation or accidental escalation may be relatively low. Nevertheless, the stakes are high and decision-makers must always be concerned about the risks, because the European theater is the most heavily militarized in the world. Both alliances have modernized and expanded their forces–conventional and nuclear–since NATO's formal endorsement of flexible response in 1967. The formidable increases in Soviet capabilities have extended the considerable Western debate over the costs and feasibility of matching Soviet conventional forces. This and the related controversy over the INF Treaty revolve on the question of whether war, once started, *can* or *should* be limited to conventional forces. This, in turn, colors perceptions about what best maintains deterrence in Europe in the first place.

One side of the debate favors a balanced deterrent posture that includes credible conventional war fighting capabilities if deterrence fails. Others fear war of any kind, and believe deterrence is best maintained by actively making the consequences of its failure so appalling that no one will let that occur. This means the explicit threat of early use of nuclear weapons. This debate, broadly described here, is a direct off-shoot of and bears remarkable similarities to the American strategic debate between mutual assured destruction (that is, MAD or large-scale attacks against urban/industrial targets) and limited nuclear war fighting options. The belief (or hope) that limited nuclear war fighting options can both strengthen deterrence and mitigate the consequences of nuclear war if deterrence fails was firmly established as the basis for US strategic doctrine by the early 1970s. The prevailing logic in that debate is equally valid for the European theater. There, the strategic counterpart of limited nuclear war is flexible response which rests on credible conventional forces with mobility, strategic reserves, and defense in *depth*.

In fairness to our German allies, it is important to emphasize that 'defense in depth' is defined here as active defenses. The alternative to forward defense is not a European landscape festooned with elaborately constructed, fixed defenses at every strategic point or a Maginot Line along the Rhine (a de facto sacrifice of nearly all of Germany). At the other extreme, a rigid forward defense posture that requires early use of nuclear weapons is the theater equivalent of mutual assured destruction. As the basis for deterrence in Europe, forward defense needs to be re-examined in light of the Soviet theater force modernization, arms control, and military strategy.

Many Western analysts have argued in some detail that the Soviets possess a credible conventional capability to penetrate NATO's forward-deployed forces with armored thrusts deep into rear areas.[18] Others also have traced a distinct shift in Soviet military writings toward a purely conventional strategy to neutralize NATO nuclear forces, including military exercises without a nuclear phase.[19]

To counter these developments while attempting to preserve both flexible response and forward defense, NATO adopted a 'deep strike' strategy or Follow-on-Forces Attack (FOFA). The new concept is heavily dependent on emerging technologies to produce *conventional* weapon systems that can acquire, track, and destroy Warsaw Pact forces 50–150 kilometers beyond the inter-German border, delaying or preventing them from reaching the battlefield.[20] Many systems are currently in research and development to meet the firepower requirements of 'deep strike'. Delivered from aircraft and ground-to-ground missiles this new generation of 'assault breakers' will dispense hundreds of submunitions on airfields, roads, key choke points, and in the path of Soviet armored columns.[21]

Deep attacks with conventional weapons may be one of the most important developments of the 1990s in reinforcing NATO's commitment to flexible response. In both declaratory policy and war planning, 'deep strike' strengthens the credibility of conventional options. The Soviets are far less likely to initiate a war in Europe that cannot be won quickly. Stalemate or protracted warfare is a threat to the political/military cohesion of the Warsaw Pact and to the long, vulnerable logistical networks that link the Soviet industrial base to its armies on the Central European front. The Warsaw Pact center of gravity or the decisive point of attack for NATO strategy is Soviet momentum on the battlefield. A capability to disrupt the tempo of large, centrally directed armies is a potent deterrent against a military doctrine that recognizes that it must win quickly or confront a rapidly deteriorating political and logistical base.

Like all NATO innovations, the new strategy is not without operational problems. The ability to see deep through the 'fog' of war will confront NATO with major challenges to deploy reliable airborne sensors to track and engage mobile targets deep in the enemies' rear.[22] Costs will also be a major factor on both sides of the Atlantic. Preliminary estimates suggest the strategy will cost $9–20 billion to implement.[23] In an era of fiscal constraints, required weapons systems will probably be fielded incrementally and face constant threats that their numbers will be insufficient to make a credible conventional deterrent.

The debates over costs and technology ignore a more basic question of strategy. New conventional weapons provide NATO with an improved defense in depth, but the depth is behind enemy lines, not NATO's. The strategy has been molded to fit the German political requirements for forward defense. Unless NATO is capable of developing a more extensive defense in depth that includes fall-back positions and larger mobile reserves, deep strikes may lack the credibility to deter war or keep it conventional if deterrence fails. 'Winning the first battle' isn't possible if NATO engages Soviet forces earmarked for the second or third battle while Soviet shock forces at the front break through NATO's thin front line. Defenses in depth would support deep strikes by wearing down or isolating Soviet armored columns from their logistical support base, centralized command structure, and from the follow-on echelons required to maintain the momentum of the attack. They are also a prudent hedge against NATO's own command and control failures that may result in limited damage to Soviet second or third echelons. Striking deep without planning defenses in depth is only half a deterrent. Europeans concerned about the future credibility of NATO need to re-examine their reluctance to give up territory for better fighting positions for at least two reasons.

First, a linear defense with a deep-strike capability provides greater incentives for a massive Soviet assault to break through and relieve the pressure being applied on Soviet rear echelons. As a supplement to echeloned assaults, Soviet conventional modernization programs have given them the capability to employ vastly improved versions of operational maneuver groups (OMGs) used successfully in World War II. These formations of mobile, high-speed armored and mechanized infantry divisions are designed to achieve rapid penetration through weak points into the operational depth of NATO, break down into smaller units, and quickly strike air and nuclear capable forces, disrupt command and control, destroy logistical bases, and seize terrain needed to facilitate rapid advances of larger units. Successful OMG operations that severely disrupt the NATO rear areas increase the chances of rapid advance without resorting to nuclear weapons.

The tactical implications of developments in Soviet deep thrust capabilities are considerable for forward defense. They require new, countervailing NATO tactics to cope with a non-linear battlefield if Soviet operational maneuver groups threaten to break through and carry out offensive operations behind NATO's forward defenses. The Army has recognized this problem if the 'deep-strike' strategy is viewed as part of a broader operational doctrine in evolution since 1979, known as Airland Battle. Airland Battle envisions mobile forces able to use quick maneuvers and decentralized execution of offensive missions to put Soviet bloc forces on the defensive. Deep strike in isolation is an elaborate, high-technology reconnaissance-strike system. Airland Battle incorporates deep strike operations as one component in a comprehensive doctrine that also includes military operations at the front or points of attack and in rear areas where enemy forces may have penetrated. Airland Battle stresses the need to plan for an integrated battlefield–deep, front, and rear. This is a sharp contrast to its doctrinal predecessor that was exclusively preoccupied with the direct battle at the front.[24]

A capacity for the early initiation of offensive operations by land as well as by air forces will be a more threatening deterrent to Soviet planners, but only if these forces can maneuver against penetrating

Soviet operational maneuver groups. The ability to redeploy forces quickly, set up for re-engagement, and perhaps be reinforced from the flanks is essential in winning battles that are likely to include retrograde movements away from or toward the flanks of Soviet mechanized forces capable of high rates of advance. It is important to underscore the relationships between forward defense and changes in Soviet concepts of offensive operations. Deep strikes against Soviet second echelons and the greater maneuverability of US forces achieved in the 1980s form the foundation for a credible conventional deterrent.

Related to the questions of maneuver and battlefield depth is the question of where the most probable avenues of attack for Soviet forces will be. Unless NATO forces were able to hold their forward positions and repel Warsaw Pact invaders, the urban/suburban growth of West Germany with its connecting transportation and road systems makes it difficult to bypass heavily populated areas. More critically, Soviet forces are very likely to aim for such areas. Expedient tactics for Soviet armored columns would be to 'dash' as quickly as possible for the cover provided by West Germany's urbanized corridors. 'Urban hugging' tactics have considerable advantages for the attacker when compared with the alternatives of fighting in open terrain or in cities. Urban hugging deployments of armor and mechanized infantry could fight through the more penetrable suburbs while avoiding both the more numerous obstacles in the inner cities and the open terrain that comprises NATO's nuclear and conventional killing zones.

Whether combat in suburbia favors the attacker or defender may depend entirely on prior planning, tactics, and weapons employed. Although official publications are optimistic, the US army is not well-versed (perhaps as a reflection of German sensitivity on the subject) or experienced in the subject of combat in built-up areas. The political sensitivity of fighting in NATO rear areas may also explain why FOFA targeting strategy was formally adopted by NATO in December 1985, but Airland Battle remains a product of American doctrinal thinkers that must be adapted to Europe in ways that are compatible with forward defense.[25]

NATO's center confronts Soviet planners with formidable obstacles. The defensibility of the terrain is augmented by scores of villages in close proximity, perhaps a mile or two apart, that can be converted into mutually supporting defensive positions. Paul Bracken has noted that 'a typical defensive position for a NATO armored brigade on the East German border contains about 85 villages'. Combined with forests, they 'comprise 60 percent of the available terrain . . . Warsaw Pact forces attacking tanks would be unable to bypass one village without almost immediately running into another'.[26]

There is growing recognition that enemy ground operations are more geographically restricted and the lethality and control of defenses are increasing. Natural terrain features and hundreds of small villages provide cover and concealment for NATO troops in the forward defense sector. West Germany's urban corridors form, if preplanned, a second echelon defense. Defending either with a strategy that requires early use of short-range nuclear weapons may leave NATO in a self-deterring posture due to fears of collateral damage and high casualty rates among non-combatants in the Federal Republic. German fears of either conventional or nuclear war are justifiable. The question, however, is what best strengthens deterrence of *both*? Official German assertions that either kind of conflict, nuclear or conventional, means total destruction has led to the sacrifice of flexible response for the sake of forward defense. The primacy of domestic politics over military strategy marks the failure to recognize that military plans and weapons employment policies are not just prudent hedges against the day when deterrence fails. They are the very core of deterrence that communicates threat and credibility to the enemy. German fears of conventional war may undermine the credibility of deterrence in future crises, making the conventional conflict they fear more likely.

Developing mobile forces and adding depth to the battlefield in Central Europe strengthens deterrence and conventional options. Contrary to Bonn's fears, too much emphasis on conventional deterrence is not likely to tempt Moscow into thinking that it could launch a limited assault. More likely, NATO's thin linear defense, Moscow's modernized strategic nuclear arsenal to deter NATO escalation against Soviet territory, and the high probability of a NATO self-deterring paralysis in using its battlefield nuclear weapons on West German territory, could convince Moscow that it could win a war in Europe quickly with conventional forces.

A bolt-out-of-the-blue attack is unlikely but in a political crisis Soviet perceptions may focus on worst case assessments of NATO capabilities and intentions rather than its ambiguous political comprises between forward defense and flexible response. Under such circumstances conventional pre-emption of NATO's nuclear forces (dual capable aircraft and storage facilities for battlefield nuclear weapons remain) may appear prudent from the east side of the inter-German border. Soviet planners should not be given more credit for clairvoyance than many of their Western counterparts in sorting out the political imperatives of *coalition deterrence* (which awkwardly accommodates both conventional war and the early use of nuclear weapons) from NATO's actual military planning for *coalition warfare*. Deterring the broad range of Soviet options may require a more credible conventional force posture which is less fettered by the political requirements of forward defense.

Strengthening conventional deterrence requires several political-military-technical remedies. Mobility of forces and greater depth to the battlefield have been mentioned repeatedly here. Within that depth, NATO forces require technical breakthroughs in conventional munitions. West Germany's natural terrain and urban growth will channel Soviet armored columns. Slowing and destroying those columns may require making every infantryman a potential tank killer and air defender. Unfortunately, the lack of an effective anti-armor weapon is still considered the most important deficiency of the US army's light infantry divisions.[27] Without cost-effective conventional weapons to attack Soviet tanks and aircraft, conventional deterrence can be neither credible to the Soviets nor popular with our allies. Strategic defense and planning for war in space seem all the more remarkable when serious doubts remain about our ability to fight on the ground.

West Germans and Europeans generally recognize that they live in a war zone if deterrence fails. There is no reason, however, to view conventional and nuclear war as equal catastrophes. Conventional war would be terribly destructive, but there are good reasons (described above) why Soviet forces would stick to the roadways through penetrable German suburbs. European cities would stand a far greater chance of survival against an enemy with no strategic bombing tradition and the memory of halting the Nazi advance in front of its own cities (Moscow, Leningrad, and Stalingrad) than they would in the less predictable horror of nuclear escalation. Above all else, Europeans should understand that American critics are attempting to strengthen deterrence and avoid all wars, and not just endorsing the relative merits of conventional over nuclear destruction. If deterrence fails, Europeans will be confronted with three alternatives–destruction and stalemate, destruction and defeat, or destruction and victory. Deterrence is essential because destruction of some magnitude is inevitable. Credible deterrence, however, requires a clear preference for destruction with victory.

In the case of the Soviets, they must be deterred not only from thinking they can win, but also from believing that victory can come quickly, before American reinforcements arrive or before military setbacks result in disarray or political instability in the Warsaw Pact's rear areas. The capability of fighting a successful conventional war is the most prudent way to accomplish that objective. The pre-war deterrent effect of a robust conventional defense in depth on Soviet decisions should not be underestimated. NATO's current reliance on early escalation and first use of nuclear weapons leaves a dangerously small margin between controlled and uncontrolled hostility.[28]

Flexible response, as it was conceived and as many Americans mistakenly believe it to be, should place the burden of escalation on the Soviets. That is the only strategy that is compatible with flexible response and strategic coupling.

Strategic coupling re-emerged during the deployment of Pershing II and ground-launched cruise missiles to Europe. The credibility of the US nuclear deterrent was perceived to have been weakened by Soviet-American nuclear parity and by Soviet theater nuclear modernization programs. Many Europeans feared that the Soviet Union no longer believed that the US would retaliate in the event of a nuclear attack on Europe. The threat of intercontinental escalation weakened the American nuclear umbrella over its NATO allies, unravelling the shared community of risk on which the ultimate credibility of the alliance rested. Intermediate-range nuclear missiles based on West German soil and capable of reaching Soviet territory, it was argued, would reinforce deterrence by guaranteeing that nuclear war would not be isolated to Europe's central front.[29]

European views of strategic coupling can best be understood in the context of escalation and thresholds which separate one level of warfare from another. These include:

- Peace to war
- Conventional war to nuclear war
- Battlefield to continental nuclear war (deep strikes to include Soviet territory)
- Continental to Intercontinental nuclear war (attacks from or against North America)
- Counterforce to countervalue[30] (military targets to urban-industrial targets)

World War II taught Europeans that once you cross the threshold between war and peace, there will be high levels of destruction at 'low levels' of conflict (namely, conventional war). Therefore, deterrence is stronger if NATO's response is less certain. Ambiguity in the form of threats to jump or escalate above more than a single threshold at a time strengthens deterrence. The mechanics of deterrence require a firm commitment from the United States to use strategic nuclear weapons in defense of Europe. This seamless web of deterrence is the essence of strategic coupling and forward defense from the European perspective.

Strategic coupling is directly related and its credibility is proportional to US strategy for fighting a limited strategic nuclear war. If deterrence fails, limited attacks against Soviet territory are viewed as a form of intra-war deterrence. The objective is to compel the Soviets to cease aggression while deterring them from further escalation. Pershing II missiles were deployed to Europe in support of this strategy. Their loss, according to General Bernard Rogers, 'removes the only certain way NATO has to strike militarily significant Soviet targets on Soviet soil without resort to general nuclear war . . . .'.[31]

Strategic coupling with either intermediate-range weapons in Europe or with limited attacks by strategic weapons is a dangerous example of a benign self-image that fails to give proper weight to an opposite, but equally plausible Soviet response. From the Soviet's view, an attack on Soviet territory is strategic regardless of the launch point. The impact zone is the critical variable. American distinctions between theater and strategic nuclear attacks against the Soviet Union have no clear Soviet counterparts. From Khrushchev to the present, attacks against Soviet territory have been viewed as strategic, not theater warfare. Intercontinental retaliation, therefore, is not viewed as escalation from the Soviet perspective. The primary mission of Soviet forces in Eastern Europe is the forward defense of Soviet territory.[32] Western ethnocentrism in determining what constitutes an escalation threshold is a potential path to uncontrolled nuclear war.

No one, including the Soviet leadership, can say with absolute certainty where Soviet escalation thresholds will actually be in any given circumstances. Their maximum tolerance of nuclear attack may be, for example, limited to the critical rail transshipment points which link Eastern Europe to the Soviet land transportation system. Attacks beyond the border, unless limited to one or two weapons in isolated areas, are likely to exceed Soviet tolerance of theater

Obstacles to Conventional Deterrence

isolated nuclear war.[33] This is one of the many uncertainties NATO must face if it loses control of its conventional options.

For Americans, strategic coupling is prudent only in the context of a well structured flexible response strategy. Strategic coupling, forward defense, and reliance on the early use of nuclear weapons are among the least obstructed paths to intercontinental nuclear war. Together they have become an escalation trap that strips away all pretense of crisis management. The US cannot, in the era of strategic nuclear parity, guarantee the security of Europe by relying on threats of escalation that would ensure its own destruction. A strategy that depends on rapid escalation to intercontinental nuclear exchanges is credible only for extreme contingencies such as massive Soviet conventional and nuclear attacks across a broad front in Europe and/or the US.[34] Soviet aggression in Europe could easily take the form of selective attacks, limited objectives, and political pressure on Western allies to opt out of NATO.

The dominant role of the extreme contingency in NATO strategy is less credible in an age of strategic nuclear parity than a strategy and force structure that can deny victory on the European battlefield. The strategy of rapid escalation to continental or intercontinental nuclear war is a strategy of punishment and self-punishment at that. A more limited, continental strategy of deterrence by denial based on conventional forces armed with high-tech conventional weapons that can attack invading forces at great distance, and, like nuclear weapons, deter the massing of Soviet forces anywhere on the European frontier is more credible than US threats of self-immolation in defense of NATO.

These arcane strategies and their interrelationships are not widely understood in the United States. Outside the defense community, few Americans could identify or describe them. This in itself makes for a weak foundation on which to build allied unity or credible deterrence. If Americans, including a majority in Congress, really understood forward defense and strategic coupling they would soundly reject them. In essence the German position has said that deterring conventional war east of the Rhine is more important than avoiding nuclear war between the United States and the Soviet Union. This attitude, widespread in government circles in Western Europe, is captured in the following propositions:

> Conventional defense preparations beyond a certain level would be detrimental to the credibility of the nuclear retaliatory threats on

which deterrence of Soviet aggression truly rests; being prepared for a conventional conflict of more than a few days in duration would imply a willingness to accept a longer conflict, which would entail unacceptable levels of destruction in Europe; while improved conventional capabilities based on emerging technologies should be pursued, partly because the United States favors them, only marginal increases in defense spending are politically tolerable or strategically necessary.[35]

Senator Sam Nunn has graphically described the consequences of these views on US efforts to maintain a credible force structure in support of NATO strategy:

> Under the Long-Term Defense Program, the United States pledged to send to Europe within 10 days a total of about six divisions ... and more than 1,500 of our first-line aircraft. Yet General Rogers' assessment was that he could not hold out conventionally long enough for these promised U.S. reinforcements to reach the European theater and make their presence felt in combat. The main reason was that most of our allies were woefully short of munitions ... Running out of ammunition in the midst of a pitched battle ... is definitely a nuclear escalator.
>
> If NATO could not fight and fight well with conventional forces for its own 30-day declared goal, we would not have a flexible response capability to match our strategy. If U.S. forces are merely a delayed trip wire connecting American nuclear might to NATO defense, the United States should recognize that and adjust accordingly ... America should not plan and pay for a robust conventional defense when our allies are planning for and paying for a trip wire strategy.[36]

In other words, nuclear deterrence needs to be subordinated to the primary task of territorial defense, and not vice versa as many Europeans prefer. This is the deadlock in coalition deterrence that the INF Treaty has brought to a head. Alliance cohesion in the future will depend on its ability to deal with Soviet conventional forces through a combination of force modernization and the newly created conventional arms control forum. What Europeans must confront, however reluctantly, is the fact that the INF Treaty reduced both the Soviet continental nuclear threat and the American ability to strike Soviet territory with nuclear weapons from Europe. This weakened strategic coupling by breaking a direct link between continental and intercontinental nuclear weapons. Proponents of coupling are mistaken, however, in their assumption that strategic coupling and extended deterrence are the same thing. Extended deterrence requires credible US efforts to assist NATO allies in their own defense, *in Europe*. Strategic coupling is neither a prudent strategy for Americans nor a reasonable expectation for Europeans.

Ironically, American Pershing II missiles and ground-launched cruise missiles may have been irrelevant checkmates to Soviet doctrinal flexibility that includes a greater willingness and capability to fight a conventional war in Europe. Europeans should not fear that the US nuclear umbrella is being removed, or that the INF Treaty amounts to a nuclear non-aggression pact between the US and the USSR. Nuclear deterrence should and no doubt will remain, but not as some Europeans prefer, as a substitute for conventional defense. Politically, the fear of conventional war in the Federal Republic will never vanish. Nevertheless, conventional defenses in greater depth offer considerably better prospects than nuclear weapons for limiting damage. The probability of containing damage outside major cities would be greatly increased. Moreover, if, as seems likely, large numbers of civilians and refugees hamper military operations, high-tech conventional weapons afford more discriminating offensive options against advancing enemy units.

Defending Europe with modern technology is no substitue for reconciling the incompatible elements in NATO strategy. Forward defense must be made less rigid before flexible response is possible, and before extended deterrence is credible. The INF Treaty and conventional arms negotiations have opened a debate in which the ability of conventional forces to deny Soviet military victory on the continent should be central. Close examination of Soviet strategy and capabilities for their own style of flexible response gives a new urgency to an old NATO problem.

NOTES

1. The persistence of US nuclear threats against a Soviet conventional attack against NATO can be seen by comparing Secretary of Defense Robert McNamara's annual report to Congress with those of Secretary of Defense Caspar Weinberger. See, for example, *The Fiscal Year 1969–73 Defense Program and the 1969 Defense Budget*. (Washington, DC: US Department of Defense, 1968), p. 81, and *Annual Report to*

Congress Fiscal Year 1988 (Washington, DC: Department of Defense, 1987), pp. 44–5.
2. Soviet 'Thresholds' are discussed in Bejamin Lambeth's, *On Thresholds in Soviet Military Thought*, P–6860 (Rand: Santa Monica, CA, 1983), pp. 4–5.
3. The West German position is clearly spelled out in the three most recent White Papers published by the Federal Minister of Defense. See *White Paper 1979. The Security of the Federal Republic of Germany*. pp. 125–7; *White Paper 1983, the Security of the Federal Republic of Germany*, pp. 142–7; and *White Paper 1985, The Situation and the Development of the Federal Armed Forces*, pp. 29 and 78. Ironically, US declaratory policy goes even further than the Germans. The Secretary of Defense's Annual Report to Congress for FY 1988 states that our strategy is to avoid specifying exactly what our response will be (p. 44). The 1985 German *White Paper* states, 'A conventional attack will initially be countered by NATO with conventional forces and means' (p. 78).
4. Milton G. Weiner, *Alternative Concepts for the Defense of NATO*, Rand Paper P–7032, October 1984, (Rand Corporation: Santa Monica, California, 1984), p. 5. See also Roger L. L. Facer, *Conventional Forces and the NATO Strategy of Flexible Response* R–3209–FF (Santa Monica, CA: RAND, 1985), pp. 15–17.
5. Quoted in Paul Bracken, *On Theatre Warfare* (Croton-on-Hudson, N.Y.: Hudson Institute, 1 July 1979), p. 21.
6. Quoted in Colonel David Jablonsky's, 'Strategy and the Operational Level of War: Part II', *Parameters*, Vol. XVIII, No. 2, Summer 1987, p. 55.
7. The author is indebted to James A. Blackwell, Jr., for many of the materials presented here on the evolution of forward defense. See his, 'In the Laps of the Gods: The Origins of NATO Forward Defense', *Parameters*, Vol. XV, No. 4, Winter 1985, pp. 64–75.
8. Quoted in the *New York Times*, 4 March 1949, p. 4.
9. Text of the agreement, *New York Times*, 20 September 1950, p. 12.
10. US Congress, Senate, *Assignment of Ground Forces of the United States to Duty in the European Area*, hearings before the Committee on Foreign Relations and the Committee on Armed Services, 82nd Congress, 1st Session (Washington: GPO, 1951), pp. 131–2, 154–5.
11. The *New York Times*, 12 October 1950, p. 24.
12. Robert C. Richardson III, 'NATO Nuclear Strategy: A Look Back', *Strategic Review*, 9 (Spring 1981), p. 41.
13. The extent of US strategic evolution toward limited nuclear war fighting options from McNamara to the Reagan administration can be seen in Scott D. Sagan 'SIOP–62. The Nuclear War Plan Briefing to President Kennedy', *International Security*, Vol. 12, No. 1, Summer 1987, pp. 22–51; and Fred Kaplan, *The Wizards of Armageddon* (New York: Simon & Schuster, 1982).
14. Described by David Schwartz, *NATO's Nuclear Dilemmas* (Washington, DC: Brookings, 1983), pp. 145–6.

15. These debates and the evolution of flexible response during the McNamara period are well known and won't be repeated in detail here. See especially Alain C. Enthoven and K. Wayne Smith, *How Much is Enough? Shaping the Defense Program 1961–1969* (New York: Harper & Row, 1971), Chapter 4; Catherine Kelleher, *Germany and the Politics of Nuclear Weapons* (New York: Columbia University Press, 1975), and David N. Schwartz, *op. cit.*, Chapter 6.
16. Enthoven & Smith, pp. 140–1.
17. *Washington Post*, 15 December 1983, p. A–14.
18. Christopher N. Donnelly, 'The Soviet Operational Maneuver Group: A New Challenge for NATO', *International Defense Review*, No. 9–82, pp. 1117–1186. See also, Peter Vigor, *Soviet Blitzkrieg Theory* (New York: St. Martins Press, 1983).
19. J. M. McConnell, 'The Soviet Shift in Emphasis from Nuclear to Conventional' (Alexandria, VA: Center for Naval Analysis, 1983), Vol. 11, p. 23FF. See also P. A. Peterson and J. G. Hines, 'The Conventional Offensive in Soviet Theatre Strategy', *Orbis*, Vol. 26, No. 3, Fall 1983, pp. 695–739, and Dennis M. Gormley, 'A New Dimension to Soviet Theatre Strategy', *Orbis*, Vol. 29, No. 3, Fall 1983, pp. 537–69. For details of Conventional War exercises see Jeff Simon, *Warsaw Pact Forces: Problems of Command and Control* (Boulder, C.: Westview, 1985), pp. 72–87.
20. FOFA was approved in 1984 at the NATO Defense Planning Council meeting. *Washington Post*, 10 November 1984, p. A–16. The depth of the battlefield is described in Benjamin F. Schemmer's 'NATO's New Strategy: Defend Forward, but Strike Deep', *Armed Forces Journal International*, November 1982, p. 68.
21. For a descriptive survey of these systems see Schemmer, Ibid, pp. 50–68.
22. The Army is developing a number of systems and adjusting its doctrine to maximize their effectiveness. For example, the Joint Surveillance Target Attack Radar System (JSTAR) will be able to locate and classify targets in the second echelon and communicate real-time intelligence directly to weapons capable of immediate engagement. The Multiple Launch Rocket System (MLRS) is one example capable of firing a variety of ground to ground missiles with submunitions that can blanket such a wide area that they are referred to as 'grid square' killers (that is, they disperse munitions over an area on the map measuring 1000 x 1000 kilometers).
23. Schemmer, *op. cit.*, p. 65. The higher range is estimated in Manfred R. Hamm's 'The Airland Battle Doctrine: NATO Strategy and Arms Control in Europe', *Comparative Strategy*, Vol. 7, No. 3, P. 197.
24. These changes are best illustrated in FM 100–5, *Operations*, Department of the Army, Washington, DC, 1 July 1976 and FM 100–5, *Operations*, Department of the Army, Washington, DC, 20 August 1982. Airland Battle provided a coherent rationale for Army programs already in production or under development, such as the M1 Tank, Bradley infantry fighting vehicle, the Multiple Launch Rocket System,

the Army Tactical Missile System, and new attack helicopters–all conventional forces.
25. Critics of FOFA and Airland Battle incorrectly saw it as an aggressive, offensive nuclear war fighting doctrine at variance with the defensive nature of NATO. This was due to the concurrent INF deployments and controversy that surrounded theater nuclear weapons. The INF Treaty and a 1986 revision of FM 100–5 reinforce Airland Battle's defensive posture. For example, deep, close, and rear operations must be closely integrated and supportive of the close (forward defense) battle. The main effort is at the front. Deep maneuvers involving division-sized forces will be the exception. See *Field Manual 100–5, Operations*, US Department of the Army, Washington, DC, 5 May 1986.
26. Paul Bracken, op. cit., p. 225.
27. This and the general problems of developing anti-tank weapons for infantrymen is discussed in the *Washington Post*, National Weekly Edition, Vol. 3, No. 14, 3 February 1986, pp. 31–2.
28. In Wintex 83, NATO's annual command and staff exercise, nuclear weapons were used six days after Warsaw Pact forces crossed the border. See Dyer, op. cit., p. 196.
29. The irony of strategic coupling is how passionately the Soviets embraced it. Soviet strategic coupling took the form of threats to attack US territory in retaliation for NATO nuclear attacks against Soviet territory. See, for example, former Minister of Defense, D. F. Ustinov's statement in *Pravda*, 7 April 1983, p. 4.
30. The author is indebted to Lawrence Freedman for these particular formulations and European perceptions of them. See 'Flexible Response and the Concept of Escalation', *Defense Yearbook, 1986* (London: Brassey's Defence Publishers, 1986), pp. 100–3.
31. Testimony before the Senate Armed Services Committee, quoted in *Army*, Vol. 38, No. 3, March 1988, p. 22.
32. The Soviets are very candid about this. A recent *Izvestia* report reminded readers that the roots of Soviet policies in Eastern Europe date back to 22 June 1941, 'when the surprise attack by Hitler's Germany and its rapid penetration of the Soviet interior . . . brutally dispelled the prewar notion that we would rout the enemy on the enemy's land . . . The group of Soviet troops in Germany is a guarantee that the past will not repeat itself'. *Current Digest of the Soviet Press*, Vol. XL, No. 14, 4 May 1988, p. 11.
33. Soviet military writings emphasize *decisive* use of nuclear weapons and denounce escalation controls and associated thresholds between levels of nuclear war. Nevertheless, they have built a complex command and control system, and a range of weapons that could support a strategy of limited nuclear war. Ironically, a Soviet declaratory strategy of limited nuclear war in Europe, even after NATO nuclear attacks on Soviet soil, would cause great anguish in Western Europe. The fact that Soviet military strategy has not exploited European fears of continental nuclear warfare supports the thesis that their primary strategy is deterrence of war from Soviet soil, and to keep war in Europe conventional for as long as possible.

34. This was the conclusion of the Department of Defense study chaired by Fred Ikle and Albert Wohlstetter, *Discriminate Deterrence* (Washington, DC: US Department of Defense, January 1988), pp. 2, 8, 30, 33–5.
35. David S. Yost, *France and Conventional Defense in Central Europe* (Boulder, C.: Westview Press, 1985), pp. 110–11.
36. Senator Sam Nunn, 'NATO Challenges and Opportunities: A Three Stroke Approach', *NATO Review*, June 1987, pp. 3–4.

2 Deterrence Soviet Style

Soviet military debates reveal the logic from which their military doctrine and strategy develop. That logic, based on the defense of Soviet territory and the acquisition of sufficient military power to deny military victory to any coalition of enemies, has not changed radically over time. The continuity in Soviet military doctrine and objectives should not, however, result in the failure to recognize the dynamic nature of Soviet military thought and military strategy. That dynamism in the European theater is the subject of this chapter which examines the evolution of deterrence–Soviet style. Its major thesis is that from Stalin to Gorbachev the political leadership has, through its military doctrine, tasked the armed forces to plan and develop military forces and strategies to deter nuclear war against Soviet territory. Military strategy has evolved with dramatic and often sudden shifts in response to American threats, Soviet capabilities, and changes in political leadership. This thesis is not new to Soviet specialists.[1] The emphasis here is on Soviet potential for flexibility in war and how its evolution differs from NATO concepts of flexible response.

The broad themes in Soviet military strategy have included: conventional war in Europe; massive, nuclear pre-emption against the United States; combined nuclear and conventional war in Europe; and more recently, an emphasis on fighting a conventional war in Europe, but under the threat of nuclear escalation. Recent doctrine and strategy have revealed a heightened appreciation of the futility of nuclear war, and a corresponding decline in the military practicality of nuclear weapons. These views coincided with the Soviet achievement of strategic nuclear parity with the United States in the 1970s. Nuclear equality brought both confidence and anxieties. Confidence was the result of unambiguous superpower status; anxieties stem from more recent Soviet unwillingness to conduct a two-front war against economic stagnation and threats from American military technology. Soviet strategic modernization programs and Gorbachev's arms control strategy serve the goal of maintaining nuclear parity, but through reductions, not races in which the Soviets prefer not to compete. The strategic posture that Gorbachev is protecting has had a long and erratic history, a history that the Soviets are not anxious to repeat.

OFFENSIVE DEFENSE: SOVIET STYLE

Soviet military strategy is more difficult to trace than its NATO counterpart. Declaratory strategies, whether conciliatory or threatening, are often designed for their propaganda value on European and American audiences. Actual operational policies or specific military strategies that the Soviets might employ in war can be gauged more confidently from careful, long-term monitoring of their military writings, force procurement programs, and military training routines.

Soviet commentaries on military strategy are limited in their value for predicting actual Soviet behavior in war. Strategic orthodoxy could easily give way to ad hoc strategies based on last-minute political and military judgments. Nevertheless, Soviet writings can, when carefully cross-checked to rule out their least plausible interpretations, serve as a guide to Soviet military preferences. It would be a mistake to dismiss Soviet military literature as a well orchestrated disinformation campaign, or as a contradictory compendium in which one can find support for almost any point of view. Such conclusions are predictably heard from non-specialists or ideological vandals who ransack available writings for supporting quotations. As Stephen Meyer has wryly observed:

> Little attention is paid to the context or the forum in which policy statements and writings appear. Articulations of Soviet foreign policy, military policy, military doctrine, military strategy and operational art are strung together without regard for the careful distinction among them made by Soviet authors. Time is judged irrelevant, as viewpoints from the 1950s, 1960s and 1970s and the present are thrown together. The product, then, is a wholly artificial construct of a 'Soviet view'.[2]

How the Soviets view strategic nuclear war and deterrence since Brezhnev's stewardship of extensive nuclear modernization programs has been the source of considerable debate. Conservative analysts have argued repeatedly that ongoing strategic modernization programs are inseparably linked to extensive Soviet military writings on the importance of being able to fight and win a nuclear war.[3]

Others argue that strategic parity has brought Soviet doctrine more in line with Western views of deterring rather than fighting nuclear wars.[4] The most important evidence of this was the defensive caste given Soviet military doctrine by Brezhnev in the City of Tula in

January 1977.⁵ This major foreign policy address greeted the new Carter administration and followed an election campaign that had put the Nixon-Ford-Kissinger détente and arms control policies on the defensive. The 'Tula Line' rejected strategic superiority as a goal in favor of equality with the United States. Brezhnev disavowed nuclear pre-emption, claiming that no one could win a nuclear war. Like Gorbachev, he described Soviet doctrine as 'purely defensive' in character. These moderating themes were reiterated by the Party leadership and eventually by the Soviet military. Soviet military publications were also marked by a substantial decline in the volume of articles devoted to nuclear doctrine.⁶

The timing of the Tula Line suggests that Brezhnev was not only struggling with Soviet public relations and image problems in the Western media, but that he was also attempting to reconcile the often contradictory relationships between military doctrine and strategy. Doctrine and strategy have a paradoxical relationship. Deterring nuclear war has required both sides to develop credible forces and plans for fighting and, if not winning, at least assuring that potential adversaries cannot win. This paradox which so often fuels mutual perceptions of threat as well as the most sincere, benign self-images can be reconciled only by a broad understanding of how military doctrine in general relates to military strategy.

Soviet *military doctrine* is the official views of the Party leadership on the general nature of war and its political/military goals. In the Soviet sense of the word, doctrine covers the preparation of the entire nation for war. It is much more general than military strategy or operational art, although it provides the framework within which the latter are developed. For example:

Defensive
- Deterrence/Denial of nuclear attacks against the Soviet homeland is vital.
- War in Europe should remain conventional for as long as possible.
- Nuclear escalation should be limited to the European Theater, excluding the USSR.

Soviet military doctrine communicates to the military leadership what it will be called upon to do. Soviet military doctrine and strategy consist of two aspects, the political and the military-technical. The political has traditionally been defensive in tone while the military-technical has been decidedly offensive. The significance of Gorbachev's new military thinking assessed below must be determined by its effect on the military-technical as well as the political side of Soviet military strategy.

Military strategy includes the basic concepts, force structure, and weapons employment policies to achieve the goals of military doctrine. For example:

Offensive
- Conventional pre-emption of NATO nuclear forces and air bases.
- Disrupt the NATO rear with operational maneuver groups.
- Be prepared for first decisive use of theater or battlefield nuclear weapons.
- Maintain stategic counterforce capability for intrawar deterrence of attacks on Soviet territory.

Operational art refines military strategy in the form of detailed planning, execution, and conduct of military operations. For example:

Offensive
- Co-ordination of deep strikes, air assault troops, OMGs, and their supporting arms.
- Tactical deployments to minimize NATO battlefield nuclear weapons.
- Co-ordination of major combined arms operations.[7]

American-Soviet strategic doctrine and military strategy have evolved from this paradoxical offensive-defensive relationship in a similar fashion, but with important differences in strategy. American doctrine rests on the deterrence of war, but military strategy has increasingly embraced 'damage limitation' (to the US) through limited nuclear attacks against Soviet strategic nuclear, theater nuclear, or conventional forces before they can be launched or fully deployed. Limited strategic attacks carry explicit threats of escalation to urban/industrial targets if aggression continues.

Soviet nuclear strategy has placed even greater emphasis on warfighting and damage limitation (to the Soviet Union) through large-scale, pre-emptive attacks against military targets. These threatening features of strategy are not (anymore than the US version) in themselves a rejection of deterrence as a doctrine. Deterrence enjoys little credibility if you do not tell your enemy what the costs of its failure will be.

Soviet doctrine and strategy reject American efforts to incorporate limited strategic nuclear war and escalation controls, emphasizing instead that the better their armed forces are prepared to fight a nuclear war, the better their society is equipped to survive its effects, and the more clearly the adversary understands this, the more he will be effectively deterred. This doctrine seeks to deny the opponent the prospect of military victory.[8] It covers all of the Soviet's strategic bases since it rests on well established war fighting strategies and capabilities in the event deterrence fails.

Both Moscow and Washington operate on doctrine that underscores their fears of nuclear war, even as their military forces prepare to conduct it on a massive scale. This irony is missed by American analysts who portray a Soviet leadership dogmatically in pursuit of a nuclear war-winning capability. Such assertions are almost always supported by research that compares US strategic doctrine (defensive) with Soviet military strategy (offensive). A substantially different conclusion can be reached when American doctrinal statements which emphasize deterrence are compared against Soviet doctrine, and American strategy is cross-checked against its Soviet counterpart. Remarkable similarities emerge. Discussions of strategy address what the military must do if deterrence fails. Specific plans and credible forces reinforce deterrence, making its failure less likely. When Soviet military commentators discuss 'victory' in a nuclear war, they make it clear that fighting nuclear wars is not the preferred course of action and that one should not always count on victory. Soviet statements have been no more specific or threatening than American formulas for 'terminating nuclear conflicts on terms favorable to the US'.[9]

Brezhnev's 'Tula Line' and the more moderate tone in Soviet military writings since 1977 are most likely, as Benjamin Lambeth has concluded, the result of Soviet embarrassment over their own strategic hyperbole. American hard-liners attacking SALT and détente drew their strength directly from Soviet military sources.[10] Brezhnev and others who echoed the more moderate line did not fundamentally alter Soviet doctrine. In Soviet eyes, doctrine had always been defensive, because war was always perceived to be thrust upon them by the West. The Tula Line and the moderation in official statements, including the 1982 formal nuclear no-first-use pledge, was a concerted effort to focus the debate on doctrine, whose primary concern was deterrence, and away from the more threatening details of military strategy.[11] These details were refined in terms theoretically different from Western concepts of deterrence and how war should be fought if deterrence fails. Like its Western counterpart, however, Soviet military strategy has also attempted to escape from its nuclear dependency.

STALIN'S CONVENTIONAL EMPHASIS: BACK TO THE FUTURE?

Flexibility and escalation were not topics of discussion in Soviet military writings during the Stalin period. Such concepts would have been ludicrous in light of the problems faced by Soviet military strategists and forces during the period of US atomic monopoly and strategic superiority. 'Victory' in war, according to Stalin's five 'permanently operating factors' would be gained by the side that was superior in: (1) stability of the rear; (2) morale of the army; (3) quality and quantity of divisions; (4) the level of armaments; and (5) organizational and leadership qualities of the commanders. Nuclear weapons were downgraded as 'terror' weapons, not capable of producing a decisive outcome in war.

The early Stalinist line was a clever, but predictable posture for a weaker power seeking to avoid atomic attack or political intimidation. Weaker powers must minimize the political-military value of any weapon or military capability that the enemy holds as a monopoly. That's what the Soviets (and later the Chinese) did while Stalin invested heavily and apparently with desperation in developing Soviet capabilities in the systems he had declared were incapable of determining the course and outcome of war.

It is against this background of Soviet weakness that Stalin sought the means of deterring a US attack while building forces that would reduce Soviet vulnerability. The Soviets were vulnerable to a direct US nuclear attack against which they could not respond in kind. On the other hand, the US had no credible conventional capabilities against Soviet ground forces which could threaten Europe. In short, as Thomas Wolfe has pointed out, 'Lacking any means to adopt a policy of nuclear deterrence, Stalin had little choice but to make the threat of Soviet land power against Europe the counterpart of U.S. nuclear power'.[12]

Stalin's military doctrine and its dependence on a large conventional force structure should not be interpreted solely as a response to a growing American strategic threat. Soviet manpower requirements were determined by a combination of threat assessments made during the late 1940s and early 1950s. The saliency or rank ordering of various threats in Stalin's calculations is not clear, but certainly they included US atomic threats, the forward deployment of US ground and air forces under NATO, the Korean war, the defection of Marshal Tito in 1948, and would-be Tito imitators elsewhere in Eastern Europe.[13]

Stalin's flexing of Soviet manpower was a reaction to these events. His military power appears to have been more than adequate to meet threats in Eastern Europe, but the Soviet army alone was not viewed by Moscow to be an adequate deterrent against growing American military might. Stalin was pushing as early as 1947 for the development of an intercontinental bomber that could threaten the United States,[14] and expressed a desire for aircraft that would 'fly higher, farther, and faster than all others'.[15] Soviet submarine-launched ballistic missile programs began under Stalin, perhaps as early as 1949–50.[16] Research efforts in land-based missiles (rocketry in this early period) by Soviet and German technicians produced design plans for developing medium and intercontinental-range ballistic missiles at about the same time Stalin tested his first atomic device in 1949.[17]

Stalin's costly programs led to the current Soviet strategic nuclear triad and theater nuclear force structure and prove that Stalin's public expressions implying a lack of appreciation for the importance of nuclear weapons hid his grave doubts that conventional forces, no matter how large, could continue to form the basis of Soviet strategic power. The real party line began to emerge only after the first successful Soviet atomic tests in 1949–tests which 'put an end to the atomic blackmail of the imperialists'[18] By 1951 Stalin's appreciation of new Soviet capabilities was fully expressed in an 'interview' with a *Pravda* correspondent:

> Atomic bombs of various calibers will continue to be tested in the future in accordance with *the plan* for the defense of our country against an attack by the Anglo-American aggressive bloc . . . in the event the U.S.A. attacks our country, the ruling circles of the U.S.A. will use the atomic bomb. It is precisely this circumstance

which has compelled the Soviet Union to have atomic weapons, in order *to be fully prepared to meet an aggressor*.[19] (emphasis added)

The inability of conventional forces to deter atomic attack was expressed by Stalin in far more colorful language to his subordinates. For example, he reportedly told his chief of aerodynamics research, Colonel G. A. Tokaev, that the Soviets needed more effective constraints for that 'gentleman shopkeeper, Harry Truman'.[20]

Soviet military literature continued to follow Stalin's doctrinal prescriptions for a conventional war fighting forces structure. Soviet writers hewing to the Party line could have been due to more than a prudent lack of assertiveness under Stalin. The debate on military doctrine which took place after Stalin's death provides evidence that many senior officers had yet to absorb the implications of strategic nuclear weapons. Their focus on operations against front line forces reflected not only an institutional tendency to cling to strategies that had worked in World War II, but also stemmed from widespread ignorance of Soviet nuclear programs. Even Khrushchev (like Truman under Roosevelt) reported that he, as a member of the Presidium (Politburo), did not know the details of Stalin's nuclear program.[21] Those few military professionals who may have had access to highly classified Soviet research and production data (it seems unlikely that anyone writing unclassified articles about Soviet military strategy would have had such access) could easily have credited nuclear weapons with relatively little significance because their numbers were small and the Soviet ability to deliver them uncertain.

The deployment of Soviet ground forces in Europe against the American nuclear threat which first posed itself from bomber bases located along the periphery of the Soviet Union gave early Soviet strategic thinking a distinct theater orientation. Soviet priorities in Europe were reflected in their initial nuclear capabilities. These took the form of regional systems – medium-range bombers and medium- and intermediate-range ballistic missiles (MRBMs and IRBMs). Soviet intercontinental capabilities against the United States were deployed only after the Strategic Air Command moved to home basing for its vulnerable, forward-deployed bombers, and as capable Soviet ICBMs became available years after their theater nuclear forces.[22]

The fact that US nuclear weapons were first deployed in Europe, and later home-based did not, in Soviet thinking, change the nature of the threat against Soviet territory. Consequently, theater and intercontinental nuclear warfare have not been as sharply separated in Soviet thinking as they have been in American commentaries. The critical difference lies in perceptions of escalation. For the Soviets, attacking American nuclear forces in Europe or in the United States was an appropriate response to an integrated, but more diffuse threat, not an escalatory step from one theater to another. By contrast, American strategic nuclear forces not only provide deterrence against direct nuclear attack against the US, but also are relied upon by Europeans for extended deterrence against nuclear or massive conventional attacks in Europe. Extended deterrence and strategic coupling spawned Western notions of flexible response and escalation control since clear distinctions were made between war in Europe and war that involved US territory. These two theaters involved clear thresholds which marked major escalation of war. Ironically, as Soviet theater and intercontinental nuclear capabilities developed after Stalin, Europeans worried that Soviet leaders might doubt the American commitment to its allies. Soviet leaders expressed equal anxiety that the US might make the mistake of thinking that it could conduct a nuclear war in Europe without risking a nuclear strike on its own territory. These anxieties were more than mutual efforts to give credibility to respective deterrent capabilities of the two sides. They also represented fundamental differences in perceptions of flexibility and escalation.

THE LEGACY OF STALIN

Stalin's death and continued progress in Soviet nuclear programs resulted in the gradual shift away from the dependency on a conventional force structure in Soviet military strategy. Debates on military strategy were accompanied by the first detailed and unclassified accounts on the effects of nuclear weapons. The Ministry of Defense publication *Krasnaya Zvezda (Red Star)* led the way in April 1954, with a series on 'The ABCs of Atomic Energy'.[23] At the same time, the military and general public alike were made aware through the popular press that a ring of hostile bomber bases was being constructed near Soviet borders, and that the American Secretary of State (Dulles) was threatening massive retaliation ('attack' in Soviet terms) in the event of war.

Confronted with the realities of nuclear weapons' effects and evidence of the American strategic threat, Soviet political and military commentaries began to question old dogmas. The process began as early as 1953 when General Talensky, the editor of *Military Thought* (major organ through which the Ministry of Defense disseminated military doctrine to officers), invited discussion of 'the laws of military science'.[24] Talensky's most serious challenge to Stalin's military dogma came in the importance he placed on surprise. 'Decisive defeat in a limited time' was possible 'given the existence of certain conditions (i.e. nuclear weapons)'. Other discussions took the more heretical position that the deployment of nuclear weapons might put the Soviet Union in a position to prevent war, and to 'paralyze the action of Lenin's Law (i.e. the inevitability of war)'. More authoritative was the election address of Premier Malenkov in March 1954 which included the assertion that 'a new world holocaust. . . . with the present means of warfare, means the destruction of world civilization'.[25] These views were rejected, however, by a still strong Party faction which restated the dogma that war would result in the destruction of capitalism, but not socialism.

These echoes of internal party debates focused on the question of how the Soviet Union should adapt to the threat posed by vastly superior American strategic forces. Malenkov's incipient notion of deterrence was premature. Within two years, however, the first Soviet bombers capable of reaching the United States would be operational in small numbers, and Soviet ICBM technology would be close to demonstrating dramatic success with Sputnik. These developments, domestic priorities, and other possible factors that the Soviets have never made clear contributed to Khrushchev's dramatic innovation at the Twentieth Party Congress in 1956 where he proclaimed the first major doctrinal change in the Soviet theory of war. War was no longer 'fatalistically inevitable'. Because of nuclear weapons, co-existence could be a permanent strategy rather than a short-term tactic. Its permanency rested on Soviet ability to deter capitalist aggression. Khrushchev's highly qualified (by his support of Wars of National Liberation) formulation opened the door for the legitimate discussion of deterrence strategies in Soviet military writings.

The debate over military strategy took a dramatic turn during the year prior to Khrushchev's speech. Marshall Sokolovsky, Chief of the General Staff, published a critical review of recent literature on military strategy, implying that it was overly tied to the past (that is, to Stalinist dogma). In *Military Thought*, Marshal Rotmistrov went further than Talensky in declaring, 'surprise [with atomic and hydrogen bombs], successfully accomplished, not only influences the course of battles and operations but in certain circumstances can influence to a significant extent the cause and even the outcome of the whole war'. He concluded with a prescient passage on emerging Soviet strategic nuclear doctrine:

> The duty of the Soviet armed forces is not to permit an enemy surprise attack on our country and, in the event of an attempt to accomplish one, not only to repel the attack successfully but also to deal the enemy counterblows, or even pre-emptive surprise blows of terrible destructive force.[26]

The editors of *Military Thought* (who often referee the debates published in their pages) condemned themselves for delaying publication of Rotmistrov's article, and also re-emphasized the importance of being ready for 'pre-emptive actions'. The danger of nuclear surprise attack and the corollary proposition that the Soviet Union needed forces in readiness, early warning, and a capability for pre-emptive attacks were all themes which were repeated by Deputy Defense Minister Marshal Vasilevsky, in the pages of *Red Star*, and even in the *Literary Gazette*.[27]

The post-Stalin turnaround in military doctrine reached its climax in a long *Military Thought* editorial published in May 1955. The task of the armed forces was described as 'above all . . . working out the ways and means of preventing surprise attack by the enemy and inflicting on the opponent preemptive blows *in all dimensions–strategic, operational and tactical*'.[28] (Emphasis added. 'All dimensions' is a reference to nuclear and conventional forces.)

Decisive action through a combined arms force structure in Soviet military strategy set the stage for major internal and external debates over Soviet intentions and capabilities. Internally, Khrushchev attempted to reverse the combined arms strategy and force structure in favor of a Stalinist-like preference for a less flexible, single-force dominant military structure. In Khrushchev's case, the Strategic Rocket Force was substituted for the Soviet army.

Externally, the consistent military emphasis on pre-emption and Krushchev's missile rattling were read by many in the West as a Soviet rejection of deterrence, and as a doctrine calling for a paralyzing first strike. From the Soviet's vantage point their doctrine must certainly have been defensive at the time. Once they accepted the proposition that strategic nuclear attack could have a decisive effect in war, the Soviets had to confront their vastly inferior capacity for launching such an attack outside the European theater. Their problem was how to minimize the crippling effects of an attack by superior American strategic forces. Pre-emptive, damage-limiting solutions and force structures aimed at denying the enemy a military victory were in sharp contrast to American declaratory policies of inflicting massive, retaliatory punishment against a Soviet attack.

The early Soviet concern with 'damage limitation' strategies is consistent with their decision to de-emphasize strategic bombers in favor of ICBMs. The bomber was well suited to Western concepts of deterrence. Khrushchev's threats could be made good if only a few aircraft penetrated American defenses. It was not, however, a reliable weapon suited to the 'defensive' mission of minimizing the harm an enemy might inflict with a massive first-strike.[29] It was Soviet concern with damage limitation and its legacy today under a massively expanded ICBM force that anchors vastly different Soviet-American views on escalation, intensity, and flexibility in the conduct of war–strategic or theater. It can be argued that ICBMs represented (with air defenses) the initial Soviet style strategic defense. In a damage limiting strategy, missiles, like artillery, could reach their targets in a hurry and could not be stopped. These virtues so attractive to Soviet strategists, even though they initially had very little to pre-empt with, have become increasingly threatening to American planners who correctly point to the potential for instability from the Soviet dual capable (offensive first-strike or 'defensive' pre-emption) strategic missile force. These concerns have been expressed by American START negotiators in Geneva in the form of proposals which distinguish between 'fast-flyers' or destabilizing ICBMs that need to be reduced in numbers and 'slow-flyers' or less threatening strategic bombers.

KRUSCHEV: STRATEGIC BLUFF OR PERIPHERAL STRATEGY?

The period from 1957, when the Soviet Union's first ICBM and Sputnik launchings startled Washington, until 1962, just before the Cuban missile crisis, was dominated by Nikita Kruschev's flamboyant attempts to profit politically from early Soviet missile successes. The bomber and missile 'gap' controversies in the United States became fortuitous allies of Khrushchev's strategic gambits which included exaggerated claims for Soviet strategic power. The USSR, Khrushchev claimed, had a five-year lead in missiles. The Soviet people could be 'calm and confident'. The Soviet's military might assured the unassailability of 'the world's strongest military power'.[30]

Khrushchev cultivated an image of Soviet strategic power that went beyond the prevailing military balance. To dismiss his actions as a strategic bluff, however, depends on how one defines 'strategic'. From a narrow American perspective, the ability to attack the respective homelands of the superpowers clearly favored the United States and its Strategic Air Command. If, on the other hand, one assumes that the greatest strategic threat to the Soviet Union came from the European theater and its array of American forward-deployed nuclear weapons, then Soviet force structure was far more capable of backing Khrushchev's boasts.

A short-term peripheral or theater strategy met the threats from American forces in Europe and a considerable portion of the intercontinental threat (although declining due to B-52 deployments), because SAC's forces, consisting predominantly of B-47s, still depended heavily on the use of forward bases (approximately ten bases) for staging and recovery. US carrier-borne aircraft and perhaps even the first few Polaris SLBMs could also be countered by Soviet theater forces. Not until 1962, as a result of completed B-52 deployments and accelerated deployment of land- and sea-based missiles would intercontinental nuclear forces represent the major element of the strategic threat to the Soviet Union.

The Strategic Rocket Forces created under Khrushchev in 1959, and on which his strategic boasting depended, began deployments of their SS-4 (MRBM) in 1959. By 1962 this force had grown to about 500–600 launchers (including over 100 SS-3s–the first MRBM) to augment a sizable Soviet medium-range bomber force (700–900 TU-4s and TU-16s), a small force of short-range SLBMs, and the first Soviet tactical missiles (FROG and Scud). These delivery systems were available to carry the estimated 1200–1700 nuclear weapons in the Soviet stockpile by 1958.[31]

Khrushchev brandished non-existent Soviet ICBMs in his speeches while he deployed a large theater nuclear force. The rhetoric and the actual force structure were compatible to the extent that Soviet theater capabilities were responsive to the bulk of the US strategic threat. Soviet ICBMs in development would come, making Khrushchev's threats good in time, hopefully before the strategic threat shifted rapidly to the American continent and the ocean's depths. In the meantime, Soviet theater forces could raise serious doubts in the minds of American strategists about their ability to retaliate massively or even moderately against targets in the Soviet Union. Until the bulk of US forces shifted to more survivable homeland bases, the issue of strategic coupling or escalation from theater to intercontinental warfare didn't exist. So long as the threat was concentrated in the European theater, Soviet strategy could remain credible through its ability to apply decisive military power in the primary theater of operation. Concerns over escalation and 'firebreaks' between theater and intercontinental nuclear war were of less immediate concern to Soviet planners than their American counterparts.

The Soviet Union entered the military competition of the 1960s with a declaratory policy of prompt escalation to nuclear war once conflict started in Europe. Soviet writers portrayed conflict in Europe *beginning* with nuclear attacks against military and urban-industrial targets, followed by conventional ground forces to consolidate gains made by initial nuclear strikes.[32] The low Soviet nuclear threshold was in sharp contrast to the emerging American concept of flexible response with its emphasis on raising the nuclear threshold, limiting collateral damage, intra-war deterrence, and conflict termination. Soviet combined arms strategy stood Western, civilian intellectual concepts of escalation on their heads, emphasizing instead, decisive military action in the early phase of war to achieve political-military objectives.

Under Khrushchev, Soviet military doctrine evolved from the reliance on a 'balanced' force structure favored by the professional military to a strategic doctrine based on the primacy of nuclear weapons and major reductions in Soviet conventional forces. The 'new look' was outlined in Khrushchev's speech to the Supreme Soviet in January 1960. In his view:

> The present level of military technology being what it is, the Air Force and the Navy have lost their former importance. These arms are being replaced and not reduced. Military aircraft is [sic] almost entirely being replaced by rockets. We have now drastically reduced, and apparently will reduce still further, or even discontinue the production of bombers and other obsolete craft. In the Navy, the submarine fleet is acquiring great importance, whereas

surface ships can no longer play the role they played in the past . . . since 1955, the numerical strength of the armed forces has been reduced by a third, but their fire-power has increased many times over during the period owing to the development and introduction of the latest types of modern military equipment.[33]

Reliance on strategic nuclear weapons in which the Soviets had a disadvantage while reducing traditional manpower-intensive forces in which the Soviets had an advantage supports the conclusion that Khrushchev was motivated by domestic economic needs and that he had confidence in the Soviet theater dominant force structure. A lack of enthusiasm from the Soviet military for Khrushchev's nuclear dominant strategy is suggested in a speech by Minister of Defense Marshal Malinovsky who addressed the Supreme Soviet the day after the Party secretary:

The main rocket troops of our Armed Forces unquestionably are the main service of the Armed Forces, but we realize that one kind of troop cannot resolve all the tasks of war . . . *the successful waging of military actions in modern war as well is possible only on the basis of the coordinated utilization of all means of armed struggle and the combined efforts of all services of the Armed Forces.*[34] (Emphasis added)

The primacy of nuclear weapons was accepted, but not the obsolescence of conventional forces in theaters adjacent to the Soviet Union. There seemed to be no disagreement, however, that war in the European theater could not occur outside the context of a general nuclear war between the superpowers. In fact, intercontinental nuclear exchanges would occur simultaneously or even precede fighting in the European theater.[35] The question of flexible response or escalation from one level of warfare to another did not exist.

In retrospect, it seems clear that the Soviet military never accepted Khrushchev's style of 'massive retaliation' based on a force structure which radically de-emphasized the role of conventional forces. The combined arms theme remained prevalent in military writings throughout the period. For example, the editors of *Red Star* summarized Soviet military doctrine under the following five concepts in May, 1962:

(1) A new world war will be a missile nuclear war.
(2) The use of nuclear-missile weapons will result in the achievement of decisive military results in a very short time over enormous areas and at any distance.
(3) *But nuclear-missile weapons have not downgraded the importance of other types of weapons. The nuclear war will be waged by massive, multimillion-man armies.* (Emphasis added)
(4) The first nuclear strike can, to a large degree, determine the entire consequent course of the war. Therefore, '... *the chief, most important, and very first priority task is to be in constant readiness for a reliable repulse of a surprise attack by the enemy and the frustration of aggressive plans.*' (Original italics)
(5) The new world war will be between two world social systems with no holds barred. But this favors the socialist side as it can be more quickly and thoroughly mobilized.[36]

The following September the first edition of Marshal Sokolovsky's widely studied *Soviet Military Strategy* acknowledged the primacy of nuclear weapons, but firmly maintained the combined arms theme:

In modern warfare, military strategy has become the strategy of missile and nuclear strikes in depth *along with the simultaneous use of all branches of the armed forces* in order to achieve complete defeat of the enemy and the destruction of his economic potential and armed forces throughout his entire territory; such war aims are to be accomplished within a short period of time ... In a missile war, the main war aims and missions will be accomplished by strategic missile forces, which will deliver massive nuclear strikes. *Ground forces in conjunction with aircraft will perform important strategic functions in modern war.* By rapid and forceful offense movements ground forces will completely annihilate the remaining army formation, occupy enemy territory, and prevent the enemy forces from entering one's own territory. *The strategic operations of other branches of the armed forces will consist of the following: the National PVO will protect the country from enemy nuclear attacks, the navy will execute military operations in naval theaters*, aimed at the destruction of enemy naval formations and naval communications, and the defense of our own communications as well as coastal areas from naval attack.[37] (Emphasis added)

However divided Khrushchev and his more 'traditionalist' marshals may have been over Soviet force structure, there was unity in Soviet strategy on the importance of early, swift, and decisive applications of military power. Absent were Soviet formulations on flexibility or escalation control in a confrontation between the two nuclear powers. Soviet concepts of limited (or limiting) warfare, when discussed at all during this period stayed close to what Thomas W. Wolfe calls the 'escalation formula'.[38] In the context of Khrushchev's peaceful co-existence formulation, limited wars (wars between states) *may* escalate into a general war, and therefore should be avoided. Limited wars between the US and Soviet Union were, by definition, not possible in the context of Soviet military strategy. Local wars or national-liberation struggles were 'just wars', involving only Soviet aid and were thus less likely to escalate into limited or general war.

Ironically, the Cuban Missile Crisis must have left grave misgivings in the minds of the Soviet military about the adequacy of their force structure and Khrushchev's declaratory policies to support both deterrence and Soviet global strategy. Cuba was a dramatic illustration of the dangers resulting from over-reliance on strategic nuclear weapons. On the one hand, the American transition from 'massive retaliation' to 'flexible response' (combined arms, but with escalation thresholds) was due to the growing Soviet nuclear prowess which robbed the American threat of what little credibility it may have had. On the other hand, the emerging US military options which ranged from counterinsurgency to limited conventional war to counterforce strategic nuclear attacks surely increased Soviet skepticism about the adequacy of Soviet force structure. It was up to Khrushchev's successors to deal with the growing threat posed by American military dexterity which threatened to rob Soviet global strategy and military doctrine of its credibility.

THE BREZHNEV PERIOD: STRATEGIC PARITY, A RETURN TO COMBINED ARMS, AND THE ORIGINS OF FLEXIBLE OPTIONS

Khrushchev's forced retirement was followed by increased discussions among the military leadership on a more balanced combined arms force structure. These discussions evolved by the late 1960s to acknowledgements that waging conventional wars or even limited nuclear wars in theaters adjacent to the USSR might be possible. Previous Soviet military doctrine had placed the conventional phase of theater warfare after an initial nuclear attack. Soviet military planners reversed that sequence, making Soviet doctrine seemingly more compatible with NATO's flexible response. Marshal Krylov, Chief of the Strategic Rocket Forces, noted:

> Military operations *will begin and will be conducted for some time with the use of merely conventional means of arms conflict.* In this case the army, the state and its economy will have some time to complete the strategic deployment of the armed forces, to take measures in mobilizing and concentrating the troops in theaters of military operations and also to reorganize industry on a war footing.[39] (Emphasis added)

Other Soviet military writers worried that no one could be certain what escalation pattern might follow an initial use of nuclear weapons in Europe.

> There is too great a risk of the destruction of one's own government . . . for an aggressor *to make an easy decision on the immediate employment of nuclear weapons from the very beginning of a war* without having used all other means for the attainment of its objectives.[40] (Emphasis added)

Just as Khrushchev had steered Soviet military doctrine away from the 'inevitability of war', Soviet doctrine in the Brehznev era reversed the inevitability of escalation to nuclear war in the European theater. These discussions, while not optimistic, represented an important advancement in Soviet doctrine. Flexibility was theoretically possible, *provided* Soviet conventional forces were decisively employed and NATO could see no incentives to escalate. These very ambitious doctrinal demands placed a heavy burden on Soviet force structure. A structure that had to look substantially different than that bequeathed by Khrushchev.

The Surge to Parity

Soviet superpower status rests squarely on its military power. That status was fully achieved during the eight years beginning with the Brezhnev regime in 1964 and ending with the signing of SALT I in

1972.[41] The Soviet strategic buildup during this period of American preoccupation with the war in Vietnam contributed to an unprecedented freedom of action by the Soviets in removing the Khrushchev legacy of strategic inferiority.

The most dramatic aspect of the Soviet buildup was their surge to parity with third generation ICBMs (see Figure 2.1). Between 1965 and 1969, Soviet ICBM forces grew at the rate of approximately 300 launchers a year, surpassing US ICBM figures in total numbers by 1969 (more than 300 SS-9s were deployed with 1030 SS-11s–the workhorse on which the Soviets rode to strategic parity). This dramatic strategic buildup was accompanied by an across-the-board expansion and modernization of ground, air, and naval forces. Together, these forces set the stage for equally dramatic shifts in Soviet strategic doctrine.

The development and deployment pattern of the SS-11 ICBM shows how important it was to the Soviet leadership to match US strategic capabilities. Routine Soviet missile acquisition had always been preceded by methodical research and development, full exploitation of existing technologies, and competition among design bureaus. The SS-11 was developed by a single design bureau, using conservative technology that was capable of fielding a large ICBM force quickly and cheaply.[42] The SS-11 appeared to be a Soviet reaction to the combined threats of the Cuban missile crisis, the trumpeting of US strategic superiority, and the earlier commitment to a large but unspecified number of Minuteman ICBMs at a rate of 250 to 300 a year.[43]

Silo construction at the first of eight SS-11 deployment areas began more than a year before its first flight test. The first missiles were operational by 1966, and the force grew steadily until 1970 when it surpassed (1030 to 1000) the US Minuteman force. Combined with the limited SS-9 and SS-13 ICBM deployments, the Soviet leadership had good reason to associate their strategic buildup with superpower equality, and the US goal in the early 1970s to stabilize arms competition through formal arms control agreements.

As Soviet military capabilities advanced under Brezhnev, declaratory policy became more cautious in contrast to Khrushchev's boastful nuclear rhetoric which often bore a notable similarity to the US doctrine of massive retaliation. Significant changes appeared in official discussions of nuclear strategy. Emphasis on the requirement for pre-emptive attacks against enemy strategic forces declined as Soviet strategic forces were deployed in more survivable,

Deterrence Soviet Style 53

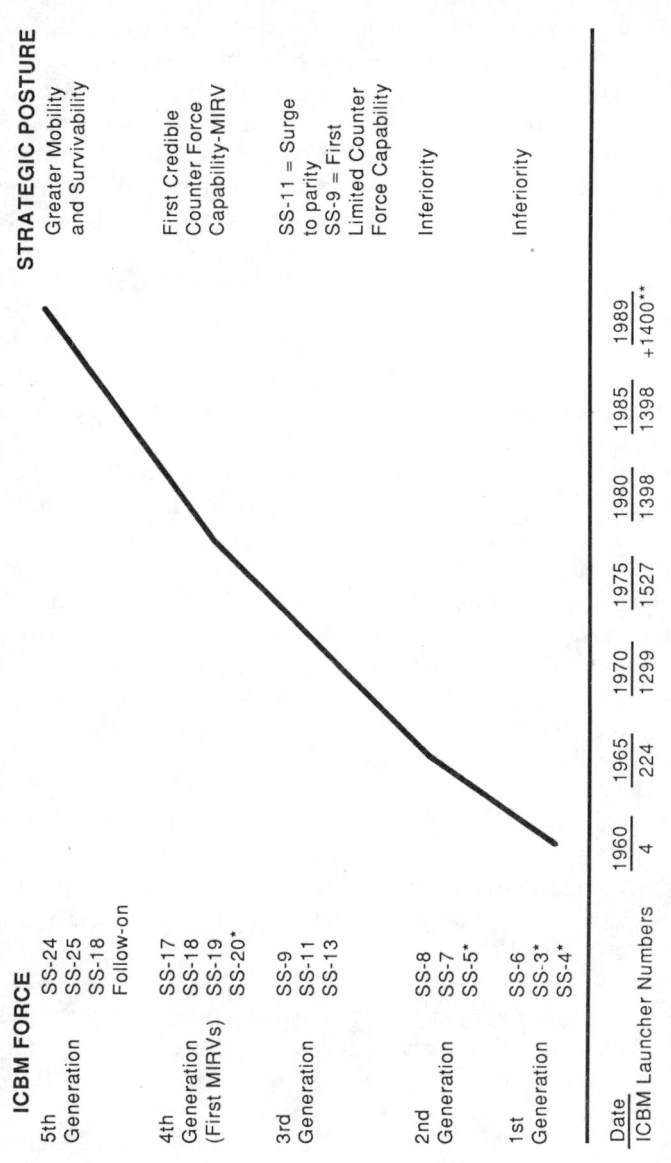

FIGURE 2.1 *Soviet strategic evolution–ICBMs*

underground silos. Soviet writings (noted above) began for the first time to suggest that a theater conflict need not 'inevitably' escalate to the nuclear level. Soviet armed forces were instructed to equip and train for contingencies solely involving non-nuclear weapons. A 1965 Warsaw Pact exercise began with a conventional phase.[44] In 1966 a classified Soviet textbook on tactics and strategy published for use at the Frunze Military Academy described the 'essence of contemporary combined-arms battle' as the conduct of combat operations with the use of conventional means of attack, while under threat of the use of nuclear weapons by the enemy.[45] Marshal Sokolovsky's third edition of *Military Strategy* (completed in 1966 and published in 1968) contained his first extended discussions of flexible response and limited war. Previous editions ridiculed the concept of limited nuclear war and the American strategy for containing war outside the continental United States. The 1968 version moved toward a more reasoned discussion of limited war being waged in a restricted geographic theater.

Soviet military writings throughout the mid- to late-1960s often seem contradictory. What appears to have taken place is a lengthy and perhaps acrimonious debate. One side firmly anchored in the Khrushchev period argued that if the nuclear powers become engaged in a war, it would inevitably escalate into general nuclear war. These views co-existed with others who argued that Soviet military doctrine did not exclude the possibility of non-nuclear warfare or warfare restricted to tactical nuclear weapons within a given theater. Others advocated preparations for limited war, both conventional and nuclear.[46]

Many of these views bore a close resemblance to the US strategy of flexible response. The co-existence of such diverse views is best explained in the context of the political succession process and subsequent rise to power of Brezhnev. Periods of succession and consolidation are nearly always accompanied by more open debate among rival political and military factions. As Brezhnev consolidated power and expanded Soviet military might, the minority voices of the mid-1960s became dominant in the 1970s. Containing wars in theaters adjacent to the Soviet Union and deterring attacks on the homeland became dominant themes in Soviet military doctrine. Perhaps the high point in Soviet attempts to shape a mutually acceptable strategy of flexible response came in Brezhnev's proposal to Henry Kissinger in 1972. The use of nuclear weapons in a war involving NATO and the Warsaw Pact should, according to the Soviet leader, be confined to the territory of allies; employment against the territory of the United States and the Soviet Union should be proscribed. Moreover, nuclear weapons should not be employed by either side in the Middle East or if there was a conventional war in any other important third country.[47]

Brezhnev attempted to bring Soviet strategy full circle, from opposition to US attempts to raise the nuclear threshold through flexible responses in Europe, to embracing mutual efforts to prevent a future war from rapid escalation to continental or intercontinental nuclear exchanges. For the Soviets, flexibility was a potential escape from the threat of nuclear escalation. For the United States, flexibility was irretrievably tied to extended deterrence and strategic coupling. In practice this confronted NATO with the dilemma of raising the nuclear threshold through conventional options while coupling those options to threats of nuclear escalation. The Soviet style of flexibility was boldly pragmatic. Strategic and theater nuclear weapons were shields meant to contain war outside the Soviet homeland. The American style had reversed this logic, offering the homeland as a shield against any level of war in Europe.

The evolving Soviet style of flexible response is by no means a carbon copy of NATO strategy. Other significant variations emerged as the Soviets responded to the evolving US prescriptions for fighting limited nuclear wars. Flexibility that raised the nuclear threshold in the European theater was acceptable, but the notion that the United States and Soviet Union could engage in limited nuclear dueling and escalation management against their respective homelands was soundly rejected.[48] For the Soviets, the critical threshold was the first nuclear weapon detonated on Soviet territory. According to their declaratory policy this would prompt immediate escalation that could not be controlled or limited.

The Soviets had good reasons for raising the nuclear threshold and exploiting their geographic and conventional force advantages in a way that might allow a future war to be fought and contained outside the Soviet homeland. To these ends, the Brezhnev era marks the beginning of an ongoing Soviet objective–an independent conventional war option. Both the force structure developed or fielded during this period, and the writings of its most prominent military figures, the late Minister of Defense, Marshal Grechko and former Chief of the General Staff, Marshal N. V. Ogarkov, support this conclusion. Grechko asserted in his 1975 book on the Soviet armed forces:

Due to a qualitative improvement in conventional means of destruction and the increase in units and formations, there has been a great improvement in the fire, shock and maneuver capabilities of troops, which permits assigning them very decisive missions on the battlefield which they are capable of accomplishing without resorting to nuclear weapons.[49]

Marshal Ogarkov, throughout his prolific writings, has consistently contrasted the stability of conventional conflict with the instability of nuclear warfare.

FLEXIBLE RESPONSE IN SOVIET FORCE STRUCTURE SINCE BREZHNEV

The evolutionary process that began in the mid-1960s with the Soviet search for operational flexibility continued throughout the Brezhnev era. By 1982, the year of his death, Brezhnev's generals were able to express confidence in their ability to conduct warfare 'both with the use of nuclear weapons and with the use only of conventional means . . . '.[50] Soviet confidence in their ability to conduct conventional operations, even in the face of NATO's tactical nuclear weapons, can be judged against the force modernization that took place during the late 1960s and early 1970s. These developments have been widely discussed in Western defense literature and will be treated only briefly here.

The Soviet army, always large and in close proximity to the European theater, was expanded and modernized. The most salient aspects of Soviet conventional force modernization were the increases in the size and mobility of individual combat units, and the growing firepower of their supporting units. The largest, best equipped, and most combat ready were deployed opposite NATO. The number of personnel in a fully manned armored and motorized infantry division increased more than 20 per cent.[51]

Soviet conventional modernization programs give them the capability to employ vastly improved versions of operational maneuver groups (OMGs) used successfully in World War II. These formations of mobile, high-speed armored and mechanized infantry divisions are designed to achieve rapid penetration through weak points into the operational depth of NATO, break down into smaller units, and quickly strike air and nuclear capable forces, disrupt command and control, destroy logistical bases, and seize terrain needed to facilitate rapid advances of larger units. Successful OMG operations that severely disrupt the NATO rear areas increase the chances of rapid advance without resorting to nuclear weapons.

Flexibility for Soviet ground forces has also resulted from modernized air and air defense forces. From 1965 to 1977, the offensive firepower of aircraft in Eastern Europe grew by 90 per cent.[52] As a result, frontal aviation has great versatility in employing long-range strike aircraft, close air support, and air defense of ground forces. Refined conventional air operations alone give the Soviets a substantial substitute for previous requirements to employ nuclear weapons against high-value military targets throughout the depth of NATO's defense.

The Soviets also increased their operational flexibility in Europe by deploying nuclear capable artillery in the early 1980s. These systems had been abandoned by Khrushchev in favor of battlefield rockets and missiles. Soviet writings suggest a narrowly defined set of roles and missions for their nuclear artillery. Concentrated conventional artillery and the operational maneuver groups remain vulnerable to NATO nuclear weapons. Limited Soviet deployments of nuclear-capable artillery allows for increased firepower with greater dispersion. Together with dual capable short-range missiles and tactical aircraft, they betray Soviet military judgments that conventional firepower cannot always be counted on for the success of ground force operations.[53] Nevertheless, nuclear artillery does allow a limited Soviet nuclear response at the tactical level. This represents a dramatic change from the force structure of the 1970s when their minimum nuclear response would have been warheads of several hundred kilotons delivered by inaccurate short- or medium-range missiles.

The cumulative effects of Soviet conventional modernization and strategic nuclear parity have been greater operational flexibility. NATO use of nuclear weapons might be kept limited through Soviet conventional suppression and Soviet initial nuclear responses would not have to be massive. Growing Soviet confidence in non-nuclear options was authoritatively expressed by Colonel-General M. A. Gareyev, Chief of the Military Science Directorate of the Soviet General Staff. Gareyev stated categorically that the assumptions and scenarios of early massive use of nuclear weapons found in Marshal Sokolovsky's *Military Strategy* had not stood the test of time. He explained that the destructive potential of nuclear weapons had become so great that their 'massive use' could lead to 'catastrophic consequences' for both sides.[54]

Soviet development of more flexible war fighting responses in the European theater created major concerns and even doubts in the minds of some Soviet military commanders. There was also considerable skepticism in the United States concerning Soviet intentions. Soviet concerns focused on their ability to maintain the conventional option against a growing array of Western high-tech weapons. American concerns focused on bellicose statements by the Soviet military on their ability to fight and perhaps even 'win' a strategic nuclear war.

Soviet Concerns About Conventional Options

General Gareyev's preference for conventional options co-existed with extensive warnings from Chief of the Soviet General Staff, Marshal Ogarkov. Ogarkov voiced alarm that Soviet military posture could be undermined by high-tech modernization of US and NATO conventional forces. New conventional means of warfare with 'precision weapons' (now called advanced conventional munitions or ACMs), and weapons based on 'new physical principles' would bring new, incomparably more destructive weapons.[55]

Ogarkov repeatedly linked new technology and military strategy. As early as 1977, he stressed that the incorporation of modern weapons and technology 'invariably entails changes in military art, in strategy . . . and the forms and methods of combat action'.[56] These themes were repeated in his writings throughout the early 1980s. It would be a mistake, however, to judge Ogarkov and his controversial reassignment in 1984 as a dispute between the military and the party over basic issues of military strategy. Ogarkov's prolific writings grew out of the broad political context of a complex period lasting from 1977 to the mid-1980s. Policy struggles during this time suggest that the Soviet leadership suffered considerable internal turmoil over the issues of maintaining their military posture (expanded at great cost since the mid-1960s) in the face of Western technology, their own economic stagnation, and the emergence of a hostile American president determined to modernize both strategic and conventional military forces.

Marshal Ogarkov's arguments were undoubtedly affected by three significant events that occurred in the 1976-77 period: the death of

Defense Minister Grechko, a dramatic decline in the growth rates of Soviet defense spending, and Brezhnev's 'Tula' speech on Soviet strategic doctrine (discussed above).

The Soviet military buildup during the decade 1966-76 was financed by a steady 4–5 per cent annual rate of growth in the defense budget. By the mid-1970s the entire Soviet economy was plagued by sluggish growth rates. Achieving strategic nuclear parity with the United States had been costly, and the Soviets faced the sobering prospect that their economy could no longer support current rates of spending. Against this background, Brezhnev appointed Dimitri Ustinov to be the first civilian Minister of Defense since Trotsky. With strategic parity secured and détente at its peak, Ustinov presided over a substantial slowdown in defense spending. The defense budget leveled off to a 2 per cent growth rate with the rate of spending on weapons procurement remaining flat well into the 1980s.[57]

Ogarkov's writings, especially his books, *Always Ready to Defend the Fatherland* (1982), and *History Teaches Vigilance* (1985–published after his removal as Chief of the General Staff in 1984), are predictable pleas from generals confronted with flat procurement budgets. For Ogarkov, the situation confronting Soviet military planners must have looked distressing. Soviet-American relations were steadily growing worse, the Reagan defense buildup was well underway, and Western technology was shaping a new NATO strategy (deep attack against Soviet second and third echelons in Eastern Europe) based on advanced conventional munitions–all this at a time when Soviet doctrine was showing a strong preference for conventional war in Europe. Many traditional tenets of Soviet military strategy were under threat: Ogarkov feared that the inherent advantage of the attacker in gaining the initiative over the defender was being reversed.

The real thrust of Ogarkov's arguments was aimed at increasing defense spending. He appeared to be appealing forcefully to the political leadership, arguing that procurement strategy and military doctrine were moving in opposite directions. By contrast, Defense Minister Ustinov argued in a major article published in *Pravda* that the armed forces had been given 'everything they needed to administer a timely and appropriate rebuff to any aggressor'.[58] He also reminded the armed forces that the major challenges facing the country meant that 'comparatively smaller increases in material expenditures and labor resources' could be devoted to the military.[59]

The reasons behind Ogarkov's dismissal as Chief of Staff cannot be stated with certainty. Rumors circulated to Western journalists in Moscow portraying him as too forceful a personality for the Party leadership caught up in the second of three succession struggles they were to confront between 1982 and 1985.[60] Ustinov could have removed him for his persistent demands for greater investments in technology. Reassigning him to an operational command effectively removed him from the resource allocation process within the Defense Council (he retired in 1988). Whatever the reason for his dismissal, Ogarkov's views spanned three succession struggles, and continued to influence the course of Soviet military strategy long after his fall from the top. The Party-military debate never revealed significant differences over threats, objectives, or military strategy. The basic issues were economic. The post-Brezhnev leadership (Andropov and Gorbachev, but not Chernenko) sought to revitalize the economy as a whole. Cuts in military spending were required for both guns and butter in the future. Ogarkov favored immediate increases in defense spending to meet the threat from Western military technology to Soviet conventional warfare options.

Gorbachev has chartered a course that seems designed to meet long-term threats of concern to both the Party and military leadership. Only by restructuring the economy and making it more productive can the Soviet system preserve its 'superpower' status. This theme has been struck repeatedly by Gorbachev. If the economy fails to grow at a rate of at least 4 per cent a year, 'the fate of socialism in the world' will be affected.[61]

Arms control has become a major component in long-term Soviet economic goals and in near-term military strategy. On the surface, arms control appears to be a strategy that bridges Party-military differences over resource allocation. Limiting the US ability to exploit the full range of its military technology is an alternative to Ogarkov's plea for greater Soviet investments to meet the same challenges. Gorbachev's concessions, which made the INF Treaty possible, and his refusal to make concessions in the START negotiations that would permit extensive testing or deployment of strategic defense systems are consistent with both Soviet economic priorities and their preferred military doctrine. The Soviets recognize that arms control (SALT I) formalized their long sought goal of strategic parity with the United States. Enlarging the arms control regime is the only way they can sustain that posture.

THE SHRINKING SOVIET NUCLEAR UMBRELLA: COVER THYSELF

Conventional options in a Soviet style flexible response are dependent on the Soviet strategic nuclear umbrella–an umbrella held exclusively over Soviet territory since the Brezhnev era. If strategic parity can be maintained, and the Soviet's ability to launch a large-scale nuclear attack against the United States remains an effective deterrent, the Soviet leadership has every reason to believe that their persistent threats of intercontinental retaliation in response to American nuclear attacks against their territory from any point will remain credible. This was a constant theme in their protracted efforts to halt US deployments of Pershing II missiles in Germany.

Soviet SS-20 missiles targeted against Western Europe were deployed to more credibly deter NATO use of nuclear weapons. The US countermove with its own intermediate-range theater nuclear forces may not have robbed the Soviets of escalation dominance, but they did increase the penalties of war through their capability to strike the Soviet homeland. The willingness to accept the American zero INF proposal is fully compatible with a Soviet desire to maintain their conventional war options in Europe while relying on strategic nuclear forces for intra-war deterrence of threats to Soviet territory. Destruction of intermediate range nuclear forces removes a dangerous stepping stone on a nuclear path strewn with NATO threats of first use.[62]

Glasnost and *perestroika* constitute a rejection of Brezhnev's domestic policy and its legacy of economic stagnation and, simultaneously, an effort to maintain his hard-won strategic relationship with the United States. Brezhnev presided over the building of strategic nuclear parity and operational flexibility. Soviet military planners were tasked to devise a military strategy for this force structure that gave high priority to destroying NATO theater nuclear forces, while simultaneously retaining the benefits of remaining in the conventional phase of war. Preventive destruction of enemy theater nuclear forces became a primary strategic mission of Soviet conventional forces. Soviet strategic nuclear forces were assigned the mission of deterring enemy long-range theater nuclear forces (before the INF Treaty) and NATO strategic nuclear forces. Conventional and strategic nuclear forces have become sword and shield–the strategic shield protects the Soviet homeland; the conventional sword gives Moscow unprecedented flexibility in the European theater. Gorbachev's policies since 1985 are consistent with efforts to maintain Soviet capabilities in the face of Western challenges–most notably, the deployment of American long-range theater nuclear forces in Europe.

Soviet officials have publicly acknowledged that deployment of the SS-20 missile was, in hindsight, a strategic error. These statements go beyond the typical pattern of a newly consolidated leadership criticizing its predecessors. Deputy Foreign Minister Alexander Bessemtnykh claimed, for example, that available technology rather than political and military analysis led to the decision to modernize the aging SS-4 and SS-5 intermediate-range missiles. Brezhnev's error, it is argued, was his acquiescence in the military's business-as-usual modernization of its ageing missiles without recognizing the risk that the 'growing Soviet threat' might provoke NATO into a military buildup that jeopardized Soviet security.[63]

Developing weapons because they are technically feasible rather than on the basis of strategic requirements is not unique to the Soviets. If Bessemtnykh's (and others) assessment is correct, the analytical failure he refers to centers on the question of how the SS-20s were coupled with Soviet nuclear strategy to deter war on the Soviet homeland, and to their conventional war fighting options in the NATO theater. Development of Pershing II missiles and ground-launched cruise missiles in Europe threatened both elements of Soviet strategy. On the one hand, Soviet theater nuclear weapons were their greatest source of fire-power. On the other hand, these same weapons in enemy hands were a threat to the initial conventional phase of war.[64] Skipping the conventional phase of war could not guarantee the destruction of all enemy theater nuclear weapons, especially the more survivable Pershing IIs and GLCMs. These weapons, in combination with others delivered by tactical aircraft or naval forces, could bring down a rain of enemy nuclear warheads on Soviet forces. The range of these forces would extend the war to Soviet territory, sparking a general nuclear war.

Soviet fears of the Pershing II missiles have always focused on its range, accuracy, and short flight time to Soviet territory. These threatening characteristics could disrupt Soviet command and control capabilities, and serve as a first salvo in an escalating intercontinental nuclear war. Intermediate-range nuclear weapons may have filled a 'gap' in NATO's deterrence strategy, but from the Soviet point of view, they represented a potential escalation trap.[65] Coupling them to war in Central Europe threatened the nuclear and conventional components of Soviet military strategy. The Zero INF proposals accepted by Gorbachev not only reduce these threats, but strengthen conventional options through the reduced range and ballistic character of the nuclear threat which, in turn, increases Soviet (and

TABLE 2.1 *Weapons destroyed under the terms of the INF Treaty signed on 8 December 1987*

SOVIET FORCES
LRINF
- SS-20 — 405 deployed missiles with 3 warheads each
  245 non-deployed missiles
- SS-4 — 65 deployed missiles
  105 non-deployed missiles
- GLCM — 84 non-deployed missiles
- SS-5 — 6 non-deployed missiles

TOTAL LRINF missiles destroyed   910

SRINF
- SS-12   220 deployed missiles
              506 non-deployed missiles
- SS-23   167 deployed missiles
              33 non-deployed missiles

TOTAL SRINF   926 missiles
TOTAL SOVIET LRINF & SRINF   1,836 missiles

U.S. FORCES
LRINF
              120 deployed Pershing II
              124 non-deployed Pershing II
              (12 in Europe – 112 in CONUS)
- GLCM   309 deployed missiles
              136 non-deployed missiles

SRINF
- Pershing 1   170 non-deployed missiles
- Pershing 1   72 deployed missiles
                   (not in Treaty, but will be destroyed)

TOTAL U.S. LRINF & SRINF   931 missiles/warheads

Source: INF Treaty

NATO) opportunities for active defense–interceptor aircraft and surface-to-air missiles–against nuclear systems that may survive conventional pre-emption. The seriousness of Soviet concerns can be judged in part by the dramatic precedent established in the INF Treaty for asymmetrical reductions on the Soviet side (See Table 2.1). Gorbachev's gratuitous inclusion of short-range ballistic missiles for which there were no American counterparts further increased the scope of Soviet reductions (SRINF with a range of under 300 miles).

These dramatic reductions reinforce the arguments made here about the declining utility of nuclear weapons in Soviet theater military strategy. How NATO confronts the Post-INF conventional balance will be the real test of how the treaty serves Western security interests. The treaty cannot be judged in isolation. It is one treaty in a growing arms control regime. The true test lies ahead in unilateral NATO decisions to modernize their conventional forces and in the degree to which the INF Treaty can be reinforced by strategic arms reductions, ground rules for strategic defense, and conventional arms control. The truest test of Gorbachev's commitment to domestic reforms, détente with the West, and his power within the Soviet leadership will be revealed through progress in conventional arms reductions. Conventional arms control more than any other will force Gorbachev to demonstrate that his promotion of doctrinal change under the guise of 'reasonable sufficiency' is more than public relations.

'REASONABLE SUFFICIENCY': REVOLUTION OR RUSE?

'Reasonable sufficiency' as a basis for Soviet military requirements and as a goal for Soviet-American arms control negotiations was first unveiled in early 1986 in Gorbachev's speech to the Twenty-Seventh Party Congress.[66] A variety of official commentators, including the Minister of Defense, have developed in general and often vague terms what reasonable sufficiency would look like for nuclear and conventional forces. In his book *Perestroika*, Gorbachev defines reasonable sufficiency as a level necessary for 'strictly defensive purposes'.[67] As this study has attempted to demonstrate, the Soviet view of their military doctrine has always been defensive. The test for 'reasonable sufficiency' lies in its application to Soviet military strategy. Its military application at home will require dramatic changes in both nuclear and conventional strategy. Soviet officials have drawn a clear distinction between its applications to nuclear and conventional forces. In the long term, 'reasonable sufficiency' calls for the complete elimination of nuclear weapons. In the foreseeable future, it describes a condition of 'strategic stability' based on 'each side retaining the capability for a retaliatory strike, but neither side for a disarming first strike'.[68]

The strategic stability formulation has long been the major focus among Western arms control specialists. It is also consistent with Brezhnev's Tula Speech which rejected pre-emption (first-strike capabilities) as a goal of Soviet military strategy. The Soviets are borrowing American concepts for both home audiences and for diplomacy and arms control. On the home front, for example, Gennady Gerasimov praised Robert McNamara's minimum deterrence strategy (a second-strike capability) based on 400 survivable warheads each with a capacity (yield) of about one megaton.[69] Garasimov concluded, 'Now a quarter of a century later, we are returning to this not particularly complex idea for securing the possibility of a destructive retaliatory strike against an aggressor'.

Soviet acceptance at the START negotiations of 50 per cent reductions in their strategic nuclear forces, and their ongoing force modernization programs support their claims of increased interest in a survivable force structure that does not rely on pre-emption or launch on warning. The Soviets are converting a vulnerable, silo-based ICBM force to more survivable and lower yield mobile missiles (SS-24s and SS-25s). Clear evidence of Soviet movement toward a second-strike doctrine must wait for future decisions on their SS-18 ICBM force. These missiles give the Soviet Union its only reliable first-strike capability against US missile silos and control facilities. Soviet negotiators have agreed to reduce their 306 missile force by 50 per cent. What happens to the remaining force and the 1500 warheads it could deliver will be critical to future strategic stability. If the SS-18s are modernized, the Soviets would retain a considerable first-strike capability. If, on the other hand, remaining SS-18s are gradually phased out and replaced by smaller and more survivable mobile ICBMs, Soviet military strategy and capabilities may begin to look less threatening. The defensive caste of these systems will ultimately be determined by modernization programs that could give Soviet mobile ICBMs and submarine-launched ballistic missiles a first-strike capability against US land-based missiles and command and control facilities.

Other Western influences can be seen in Gorbachev's sufficiency formula. His goal as stated through a series of arms control proposals between 1985 and 1989 is a gradual movement toward minimal levels of arms. These general goals have been accompanied by challenges to the bureaucracy (*perestroika*) that bear some resemblance to Robert McNamara's whiz kids who attempted to impose a dramatic new management style on the Pentagon. *Perestroika* imposed on the Ministry of Defense emphasizes one of Gorbachev's stated goals of subjecting defense decisions to more rigid cost-benefit analysis, in which potential political and economic costs are weighed against projected military benefits.[70] Getting 'smarter about what we do' will, in practice mean bringing more civilian expertise into defense decision-making and promoting bright young officers whose operational roots are not lodged in the lessons of World War II. Gorbachev has accomplished this by encouraging civilian defense experts at the USA-Canada Institute and the International Department of the Central Committee to engage the military in debate. Unlike his predecessors, Gorbachev wants competing centers of military analysis and threat assessment. Threat assessment dominated by the military has never produced recommendations for defense cuts. Who determines how much is enough for Soviet security is directly related to the larger issue of making resources available for *perestroika* to succeed.

The START proposals agreed on by the Soviets and their strategic modernization programs noted above are fully consistent with the defensive caste of 'reasonable sufficiency' (deterrence via second-strike capability). A more difficult test for Gorbachev's declared goals and his bureaucratic struggle with the military will be in the arena of conventional arms control. Applied to conventional arms, sufficiency means 'ensuring an adequate defense potential so that the aggressor should not be able to count on a local blitzkrieg or on escalating such a conflict with impunity'.[71] Soviet military leaders use several different formulations designed to serve their interests. For example, in his first statement on military doctrine, Defense Minister Demitri Yazov spelled out the principles of reasonable sufficiency for conventional forces in greater detail than had any other Soviet leader.

For conventional means, sufficiency involves a quantity and quality of armed forces and arms capable of reliably assuring the collective defense of the socialist community. The limits of reasonable sufficiency are set not by us, but by the actions of the United States and NATO. The countries of the Warsaw Pact do not seek military superiority or aspire to greater security, but they will not accept less security or permit military superiority over themselves.[72]

Yazov ended with the open-ended argument that the decisive factor in preventing war remained 'military strategic parity'. Yazov subsequently noted that parity could be achieved through Warsaw Pact proposals to employ the sufficiency principle in East-West relations by reducing forces 'to such a level where neither of the sides, while insuring its defenses, has the forces or means enabling it to mount offensive operations'.[73] Other Soviet commentators have been more specific, emphasizing that less threatening deployments of conventional forces include removing tanks and other offensive weapons from front line areas.[74]

The goals of Gorbachev's new strategy gradually emerged during the preliminary phases of the Conventional Forces in Europe (CFE) negotiations. These multilateral negotiations cover NATO and Warsaw Pact military forces from the Atlantic to the Urals. The Warsaw Pact adopted the position taken by Gorbachev in Prague in April 1987, when he acknowledged that asymmetries existed in Europe 'due to history, geography, and other factors', and that 'we are in favor of removing disparities . . . by reducing their numbers on the side that has superiority in them'.[75] What constitutes military capabilities, much less superiority or equal security is a subject that challenges the skills of negotiators within and between the two alliances. Soviet officials proposed a three-stage approach to the negotiations:

- Identification and elimination of the existing asymmetries in the armed forces of both NATO and the Warsaw Pact (includes data exchanges and on-site inspection).
- Further reductions of the armed forces down to agreed upon levels–for example, by 500 000 men on both sides.
- Orienting remaining forces toward purely defensive structure and organization.[76]

Gorbachev pre-empted formal negotiations with dramatic unilateral cuts offered at the United Nations. He pledged to reduce Soviet armed forces by 500 000 men and 10 000 tanks within two years. Six tank divisions are to be withdrawn from East Germany, Czechoslovakia, and Hungary and disbanded by 1991, reducing Soviet forces at the central front by 50 000 men and 5000 tanks. Gorbachev added that remaining forces would be reorganized in more defensive postures.[77]

If Gorbachev is powerful enough to follow through, his doctrine of non-offensive defense will alter Soviet force structure and military strategy in revolutionary ways. Soviet military theorists have traditionally identified four basic types of action in war: (1) offensive operations against the enemy; (2) meeting engagements or encounters by two advancing forces; (3) defense; and (4) withdrawal. The fundamental tenet of Soviet military strategy has been that victory can be achieved only through offensive operations and the decisive use of military force.[78] The standard Soviet military text on *Tactics*, published as recently as January 1987, sets forth traditional Soviet doctrine.[79] This is a sharp contrast to more recent guidance given Soviet officers by Minister of Defense Demitri Yazov in his book, *In Defense of Socialism and Peace*. Yazov proclaimed, 'Soviet military doctrine considers the defense as the main form of military operations'.[80]

Gorbachev, through his Defense Minister, proposed a fundamental shift in military strategy and force structure from offensive to defensive. Military power sufficient to repel rather than pre-empt aggression calls into question the Soviet military's long-standing reliance on the strategic offensive to defeat the enemy. Radical change in the traditional roles and mission of the armed forces inevitably encountered resistance. Soviet military literature indicates disagreement over the meaning of these changes and their impact on force structure. As noted above, Soviet diplomats use the term 'reasonable sufficiency', while military leaders prefer 'defense sufficiency', 'parity', 'reliable defense', or 'equal security' to describe Soviet arms control objectives. These vague formulations allow them more flexibility in determining force levels and strategy. Former Chief of the Soviet General Staff, Marshal Sergei F. Akhromeyev, defined sufficiency as 'having forces, both in quantity and quality, commensurate with the level of military threat'. This debating style is typical of political discourse inside the Soviet Union. You embrace a concept, but redefine it in subtle ways that are consistent with your bureaucratic interests. This increases the difficulty of knowing how deeply rooted the support for Gorbachev's reforms really are.[81]

A more extreme view was expressed by General Ivan Tretyak, Chief of Air Defense Forces. 'Reliable defense' is, he argued, the key requirement and 'defense of the country must be absolute'.[82] Prior to Gorbachev's UN speech, he warned against unilateral cuts in Soviet forces. General Tretyak described Khrushchev's reductions in the army as a 'terrible blow to our defense capacity' and warned that unilateral cuts should be examined 'a thousand times over'.[83] Tretyak expressed the dominant view among senior Soviet officers by going on record opposing unilateral cuts and civilian analysts who advocated them for their alleged economic benefits to the Soviet economy. The 'resignation' of Marshal Akhromeyev was also a strong indication of military opposition to Gorbachev's unilateral initiatives toward reasonable sufficiency.

Nevertheless, Warsaw Pact communiqués that have defined conventional arms control objectives as mutual force structures that preclude a surprise attack or mounting a general offensive are revolutionary for an alliance that has emphasized large tank armies, mechanized infantry, massed artillery, air support and air defense. Large, asymmetrical reductions and redeployment of those forces would be unparalleled developments in Soviet military strategy.[84] For Soviet military literature to respond to these changes in such mild terms as 'equal security' and 'parity' is in itself, a significant achievement.

Still, military formulations of 'sufficiency' are a far cry from Gorbachev's doctrinal shift at the Twenty-Seventh Party Congress. For the first time, a Soviet leader proclaimed that in the nuclear age no state can rely on defense alone. Ensuring security more and more becomes a political task that can only be solved by political means. In his UN address, Gorbachev again challenged the traditional view that the Soviet Union's superpower status can only be maintained through military strength. 'It is now quite clear that building up military power makes no country omnipotent. What is more, one-sided reliance on military power ultimately weakens other components of national security'.[85]

Gorbachev's success will depend on Party cohesion and strong leadership to pull the military out of its deeply rooted preference for decisive offensive operations. Doing so is tantamount to saying that the revered lessons of the Great Patriotic War no longer apply in the European theater. Equal constraints on NATO forces will be essen-

TABLE 2.2 *Summary of Soviet Doctrinal Evolution*

Stalin, 1945–53:
— nuclear weapons not decisive
— stressed primacy of large conventional forces

Khrushchev, 1956–64:
— war no longer 'fatally inevitable'
— greater reliance on nuclear weapons
— strategic rocket forces (SRF) created, 1959
— de-emphasized conventional forces
— war in Europe nuclear from the start

Brezhnev, 1964–76:
— surge to ICMB parity
— conventional forces modernized
— war could begin conventionally
— protracted conventional war possible
— nuclear war still basic
— but, could be limited to the European Theater
— war possible without attack on Soviet homeland
— conventional pre-emption of NATO nuclear forces desirable
— be prepared to use nuclear weapons in all theaters if nuclear war unavoidable

Brezhnev and Ustinov, 1977 'Tula Line':
— rejected pre-emption, first strike
— rejected view that nuclear war could be 'won'
— Soviet goal is strategic parity, not superiority
— endorses mutual deterrence

Ogarkov:
— first-strike not decisive
— deterrence by denial, but
  — deny victory with a 2nd strike
  — war fighting emphasis must remain, but emphasis on 2nd strike and denial of victory to the aggressor
— 'unacceptable damage' defined as inability to conduct war and mount serious operations

Gorbachev, 1986:
— reasonable sufficiency
— non-offensive defense
— strategic stability based on 2nd-strike capability
— rejection of first-strike capability for strategic or conventional forces

---

tial to the success of future conventional arms control negotiations. Soviet demands in conventional arms control negotiations may be surmised from their military strategy. Soviet agreement on reductions is likely to require that the primary NATO threat against which conventional pre-emption has been directed, namely NATO's nuclear weapons and supporting infrastructure, is removed. Conventional arms control from the Soviet view, will require the 'triple zero' option–the elimination of battlefield nuclear weapons with ranges below 300 miles–which the United States has already rejected. A stalemate similar to the long SDI-START deadlock was apparent before negotiations began. The long road ahead will reveal Gorbachev's true intentions. The immediate question for Americans is whether the Bush Administration can match his strategic vision with one of its own.

CONCLUSIONS

Soviet military strategy has been dynamic, changing much like its American counterpart with each new level of capability. Ironically, Soviet declaratory policies (doctrine) have become more moderate as Soviet military capabilities have increased. The most important achievement noted here is Soviet achievement of strategic nuclear parity under Brezhnev. This created opportunities for greater operational flexibility, flexibility that led to two major strategic goals: deterrence of nuclear war against Soviet territory and the development of a conventional war fighting option in Europe. Gorbachev's arms control strategy supports these major goals.

'Reasonable sufficiency' is fully consistent with Soviet START and INF positions. Applied to conventional arms control, it is a no-lose strategy. Protracted negotiations may weaken NATO cohesion, head off conventional force modernization of the kind so feared in the writings of Marshal Ogarkov, or create momentum for the denuclearization of Europe. Any or all outcomes strengthen the Soviet conventional option while weakening the escalatory mechanism in NATO's strategy of flexible response.

The Gorbachev revolution in military affairs will not satisfy Western security interests until large numbers of Soviet forces have been demobilized and others redeployed deep into their own territory. Making Soviet military strategy and force structure look defensive requires the combined efforts of Western diplomatic skills and a compelling need for Mikhail Gorbachev to turn his revolution and his energies inward. There, he is finding an entrenched Party-State-Military bureaucracy that is as formidable an opponent as any Gorbachev has in the West.

NOTES

1. Several Western analysts still argue that Soviet military strategy has undergone little change. See, for example, William T. Lee and Richard F. Staar, *Soviet Military Policy Since World War II* (Stanford: Hoover Institute Press, 1986), and Dan L. and Rebecca V. Strode, 'Diplomacy and Defense in Soviet National Security Policy', *International Security*, Vol. 8, No. 2, Fall 1982.
2. Stephen Meyer, *Soviet Theatre Nuclear Forces, Part I: Development of Doctrine and Objectives*, Adelphi Paper No. 187 (London: International Institute for Strategic Studies, 1984), p. 6.
3. The most often quoted is former NSC staffer and Harvard Professor Richard Pipes, 'Why the Soviet Union Thinks It Could Fight and Win a Nuclear War', *Commentary*, July 1977, pp. 27–34
4. See, for example, the extensive writings of Raymond L. Gartoff or Lambeth, Ross, and MccGwire cited below.
5. 'Speech of Comrade L. I. Brezhnev', *Izvestia*, 19 January 1977.
6. Benjamin S. Lambeth, *Has Soviet Nuclear Strategy Changed?* P–7181 (Santa Monica, Ca.: RAND, 1985), p. 5. Lambeth also notes that previous hawkish military commentators were dismissed as 'irrelevant military pedagogues'.
7. An excellent tutorial on the distinctions between doctrine and strategy can be found in Stephen M. Meyer, *Soviet Theatre Nuclear Forces, Part I*, op. cit., pp. 3–5.
8. For discussions on the respective deterrence strategies of *denial* and *punishment* see Glenn Snyder, *Deterrence and Defense* (Princeton University Press, 1961), pp. 14–16, and Dennis Ross, 'Rethinking Soviet Strategic Policy: Inputs and Implications', *Journal of Strategic Studies*, Vol. 1, No. 1, May 1978, pp. 3–30.

9. Caspar W. Weinberger, *Annual Report to Congress Fiscal Year 1988*, 1 January 1967, p. 22. This formulation can be found in a number of US defense documents. No serious discussion has occurred on how coalitions terminate conflict on favorable terms to allies in various aggrieved states.
10. See Lambeth, *Has Soviet Nuclear Strategy Changed?* op. cit., p. 10.
11. The no-first-use pledge was made in a *Tass* communiqué on 15 June 1982. This theme was used extensively during the long debate over deployment of US theater nuclear missiles.
12. Thomas Wolfe, *Soviet Power and Europe 1945–1970* (Baltimore: Johns Hopkins Press, 1970), p. 34.
13. Stalin's Yugoslav problem is discussed in Boris Nicolaevsky's *Power and the Soviet Elite* (New York: Hoover Institute, 1965), pp. 250–51.
14. Colonel G. A. Tokaev, *Stalin Means War* (London: Weidenfeld & Nicolson, 1951), pp. 114–18. (Tokaev was a Soviet Defector).
15. K. A. Vershinin, quoted in *Pravda*, 18 July 1948.
16. This conclusion is Michael MccGwire's. See *Soviet Naval Developments* (New York: Praeger 1973), pp. 78, 153. It is often overlooked in accounts of Soviet-American competition that the Soviet Union possessed a small force of ballistic-missile-launching submarines well before the first US *Polaris* went on patrol in 1960. Soviet boats became operational between 1957 and 1959. See Barton Wright, *World Weapon Database, Vol. I, Soviet Missiles* (Lexington, M.: Lexington Books, 1986), p. 33.
17. Robert Berman and John C. Baker, *Soviet Strategic Forces* (Washington, DC: Brookings, 1982), p. 39.
18. K. E. Voroshilov speech of 9 March 1950. Quoted in *Current Digest of The Soviet Press*, Vol. II, No. 11, 29 April 1950, p. II.
19. *Pravda*, 6 October 1951, p. 1.
20. Tokaev, op. cit., p. 115.
21. Soviet nuclear research began on a modest scale in 1942, gathered momentum in January 1945, when espionage reports from Klaus Fuchs (and perhaps others) indicated that American scientists were making progress on the 'Manhattan Project'. It is unlikely that Stalin realized the full impact of the bomb until after Hiroshima. See Sergei M. Shtemenko, *The General Staff* (Moscow: 1963), p. 359. Quoted and discussed in Adam Ulam's, *Stalin* (New York: Viking Press, 1973), p. 625.
22. These priorities are discussed in Arnold Horelick's, 'The Evolution of Soviet Strategic Nuclear Thought'. Paper presented at the Conference on Psychology of US-Soviet Relations: Perceptions and Misperceptions, Bellagio, Italy, 1985, pp. 1–2.
23. These are discussed in Herbert S. Dinnerstein's *War and The Soviet Union* (New York: 1962), p. 222, and Arnold Kramish, *Atomic Energy in The Soviet Union* (Stanford University Press, 1959), pp. 124–5, 129.
24. Discussions of the Talensky debate are drawn from Dinnerstein, op. cit, pp. 36–49.
25. Ibid., pp. 67–71.

26. The Rotmistrov and Sokolovsky quotes are from Dinnerstein, op. cit., pp. 184–94.
27. Ibid., pp. 183–4.
28. A full translation of this important doctrinal statement was made by J. R. Thomas, *World-Wide Historic Victory of the Soviet People*, Rand T–110 19 January 1959.
29. Lieutenant Colonel V. Bondarenko, 'The Contemporary Revolution in Military Affairs and the Combat Readiness of the Armed Forces', *KVS*, No. 24, December 1968, p. 26.
30. Khrushchev's speech to the Supreme Soviet in January 1960. Quoted in *Pravda*, 15 January 1960.
31. See Stephen M. Meyer, *Soviet Theatre Nuclear Forces, Part II: Capabilities and Implications, Adelphi Paper 188* (London: International Institute for Strategic Studies, 1983), pp. 608.
32. V. D. Sokolovsky (ed.), *Military Strategy*, second edn (Moscow: 1962), p. 231.
33. Excerpts from Khrushchev's speech are from Harriet and William Scott, *The Soviet Art of War* (Boulder, Co.: Westview Press, 1982), pp. 163.
34. Ibid., p. 166. Malinovsky repeated this message the following year in his address to the Twenty-Second Party Congress. For a review of the tension between Khrushchev and his marshals, see Thomas W. Wolfe, *Soviet Strategy at the Crossroads* (Cambridge, Mass.: Harvard University Press, 1964), pp. 91–109.
35. Marshal V. D. Sokolovsky, *Soviet Military Strategy* (New York: Crane Russak, 1975), pp. 290–5 (Russian edn published in 1962).
36. 'Concerning Soviet Military Doctrine', *Red Star*, 11 May 1962, pp. 2–3; trans. in K. R. Whiting, '*Red Star* on Doctrine', *Air University Quarterly Review*, XIII, No. 4, Summer 1962, pp. 142–150.
37. V. D. Sokolovsky, editor, *Soviet Military Strategy* (trans. by H. S. Dinnerstein, Leon Goure, and Thomas W. Wolfe), (Englewood Cliffs, N.J.: Prentice-Hall, 1963), pp. 93–5.
38. Thomas W. Wolfe, *Soviet Military Strategy at the Crossroads* (Cambridge, Mass.: Harvard University Press, 1964), p. 118.
39. N. Krylov, 'The Nuclear Missile Shield of the Soviet State', *Military Thought*, No. 11, 1967, pp. 17–18. See also Major General N. A. Lomov, 'The Influence of Soviet Military Doctrine on the Development of the Military Art', KVS, No. 21, November 1965, p. 16, cited in Thomas W. Wolfe, *Soviet Power and Europe*, p. 451.
40. 'Soviet Military Doctrine and Strategy', *Military Thought*, No. 5, 1969, p. 49.
41. These developments are described in Benjamin Lambeth's *The Logic of Soviet Defense Policy*, forthcoming from Princeton University Press, Chapter V, manuscript p.139.
42. Ibid., pp. 163–7. Lambeth also notes that the standard Soviet practice of 'competitive prototyping' during ICBM R&D is responsible for the 'flea-market' of diverse systems (eight distinct types deployed) in the Soviet ICBM inventory. US planners would find the practice unacceptably complicated and expensive.

43. William G. Hyland has noted that the pace of Soviet ICBM production and deployment followed the same curve as US ICBMs. This similarity was obscured because actual Soviet ICBM deployments in the late 1960s exceeded CIA estimates, so the buildup was treated as a virtual blitzkrieg. See his *Mortal Rivals: Superpower Relations from Nixon to Reagan* (New York: Random House, 1987), p. 82. The US Air Force wanted a Minuteman force of 3000 ICBMs. McNamara cut this to 1000, but only after the decision to deploy MIRVs had been added to the compromise.
44. Michael MccGwire, *Military Objectives in Soviet Foreign Policy* (Washington, DC: Brookings, 1987), p. 27.
45. This textbook was edited by Major General V. G. Reznichenko, translated by the US Army Intelligence and Threat Analysis Center, Arlington, VA. The text was declassified in 1983.
46. This evolution is discussed in Dennis Gormley *et al.*, *Soviet Perceptions of and Response to U.S. Nuclear Weapon Development and Deployment*, Pacific Sierra Report 1211 (Albuquerque, New Mexico: Sandia National Laboratory, 1982), Ch. II.
47. Henry A. Kissinger, *Years of Upheaval* (Boston: Little, Brown, 1982), p. 277. Kissinger's response to the Soviet *démarche* made it clear that such an agreement would be seen as a cynical form of US-Soviet condominium, and would promote European neutrality since it would guarantee the devastation of each country's allies.
48. In a *Tass* interview (1984) Defense Minister Ustinov insisted that a nuclear attack on the USSR would inevitably lead to a swift retaliatory strike on both the territory where the missiles are located, and the territories from which the commands for their use are issued. 'There must be no doubt about this.' Quoted in Mary C. FitzGerald's *Marshal Ogarkov On Modern War: 1977–1985*, Professional Paper 443/March 1986. Center for Naval Analyses, Alexandria, VA, p. 26.
49. A. A. Grechko, *The Armed Forces of the Soviet Union* (Moscow: 1975), trans. by US Air Force, Soviet Military Thought Series, No. 12 (Washington, DC: Government Printing Office), pp. 147–8.
50. Major General M. M. Kir'yan, *Military-Technical Process and the Armed Forces of the USSR* (Moscow: 1982), p. 312. Quoted in Dennis Gormley, 'Soviet Military Assessments of and Counters to Western Strategy: A U.S. Perspective', paper presented at RAND conference, on Extended Deterrence and Arms Control, San Diego, CA, 1986, p. 17.
51. *Soviet Military Power* (Washington DC: US Department of Defense, 1981), pp. 27–8. Manpower increased from 9000 in 1966 to 11 000 by 1980 in armored divisions, and from 11 000 to 13 000 in a motorized rifle division.
52. Robert Berman, *Soviet Air Power in Transition* (Washington, DC: Brookings Institute, 1978), p. 54.
53. The evolution of nuclear artillery is discussed in Dennis Gormley *et al.*, *Soviet Perceptions of and Response to U.S. Nuclear Weapon Development and Deployment*, op. cit., pp. 135–40.
54. Colonel General M. A. Gareyev, *The Views of M. V. Frunze and*

Contemporary Military Theory (Moscow: Voyenizdat, 1985), pp. 239–41. Discussed in John G. Hines, Phillip A. Peterson, and Notra Trulock III, 'Soviet Thinking on Nuclear Weapons and War in the NATO Context', unpublished paper, p. 17.
55. N. V. Ogarkov, 'Reliable Defense to Peace', *Red Star*, 23 September 1983, p. 2 and 'The Defense of Socialism: The Experience of History and the Present', *Red Star*, 9 May 1984, p. 3.
56. N. V. Ogarkov, 'Defense of the Socialist Fatherland is a Matter for all the People', *Red Star*, 27 October 1977.
57. The CIA did not acknowledge the decline in the rate of Soviet defense growth until 1983. See *Defense Week*, 18 November 1983, pp. 6–7.
58. *Pravda*, 17 November 1981, quoted in Martin Walker, *The Waking Giant: Gorbachev's Russia* (New York: Pantheon, 1986), p. 129.
59. *Red Star*, 8 December 1982, p. 2.
60. Described by Dusko Doder in *Shadows and Whispers: Power Politics in the Kremlin from Brezhnev to Gorbachev* (New York: Random House, 1986), pp. 230–1. Konstantin Chernenko could easily have discovered any pretext to remove this Andropov loyalist.
61. Speech published in *Pravda*, 26 February 1987.
62. Several Western analysts have concluded that post-1977 Soviet military literature provides evidence of a consensus on the declining utility of nuclear weapons. The primary mission of Soviet strategic nuclear weapons is limited to deterrence of US or NATO nuclear attacks against the Soviet homeland. See especially, Michael MccGwire, *Military Objectives in Soviet Foreign Policy*, op. cit.; and Mary C. FitzGerald, *Marshal Ogarkov on Modern War: 1977–1975*, op. cit., and her 'The Strategic Revolution Behind Soviet Arms Control', *Arms Control Today*, June 1987, pp. 16–19.
63. *General Secretary Mikhail Gorbachev and the Soviet Military*, Report of the Defense Policy Panel of the Committee on Armed Services, House of Representatives, One Hundredth Congress, second session (US Government Printing Office: Washington, DC, 13 September 1988), p. 10. See also, *Washington Post*, 29 November 1987, p. 1.
64. Colonel A. A. Sidorenko, *The Offensive* (Moscow: Military Publishing House, 1970), trans. by the United States Air Force, p. 132. Sidorenko was Director of Military Science at the Frunze Military Academy. He placed great emphasis on the destruction of NATO nuclear weapons. While he does not rule out the use of nuclear weapons, his focus is clearly on conventional means – air strikes, counterbattery fire (at a time when the Soviets had no nuclear artillery), good intelligence, and finally, tank and motorized rifle units to finish off those not completely destroyed by other means. See discussion on pp. 134–7.
65. The Reagan administration stressed that deployment of GLCMs and Pershing II missiles to Europe was a response to Soviet SS-20 medium-range missiles. Military leaders, most notably General Bernard Rogers, have always maintained that US missiles filled a deterrence gap and were needed regardless of the SS-20s.
66. Speech is printed in *Daily Report, Soviet Union*, Foreign Broadcast Information Service (FBIS), 26 February 1986, p. 34.

67. Mikhail Gorbachev, *Perestroika: New Thinking for Our Country and the World* (New York: Harper & Row, 1987), Chapter 7.
68. *Washington Post*, 30 November 1987, p. A6. the Soviet term for stability, *Stabilnost*, has come to mirror image its Western counterpart in describing a strategic posture in which there are no incentives for a first strike. This is the formulation used by Soviet diplomats and more frequently in military journals. The military and technical community also, but less frequently, uses *Ustoychivost*, a form of stability with functional or operational meaning. For example, minimum susceptibility to disruption that could be achieved by pre-emption, launching on warning, or deployment of forces in survivable modes.
69. Speech on Moscow television, 24 May 1987. Quoted in *Foreign Broadcast Information Service* (FBIS) SOV-87-122, 25 June 1987, p. AA10. Gerasimov was quoting a study by the Soviet Scientists' Committee in Defense of Peace and Against the Nuclear Threat. Neither mentioned that McNamara's figure of 400 warheads was the number required to survive an attack, penetrate Soviet defenses, and inflict 50 per cent destruction on Soviet industry and 25 per cent kill rate on the Soviet population. Lieutenant General Mikhail Mil'shtein was critical of the McNamara formulation during a round-table discussion published in *New Times* on 3 July 1987, pp. 18–19.
70. Described in the *Washington Post*, 29 November 1987, p. 1. The Soviet source was Valentin Falin, Director of *Novosti* information service.
71. Vialy Zhurkin, arms specialist, writing in the weekly *New Times*. Quoted in the *Washington Post*, 29 November 1987, p. A1.
72. Yazov, 'The Military Doctrine of the Warsaw Pact, A Doctrine of Defense of Peace and Socialism'. Quoted in Raymond L. Garthoff's, 'New Thinking in Soviet Military Doctrine', *The Washington Quarterly*, Vol. ll, No. 3, Summer 1988, p. 140.
73. *Pravda*, quoted in the *Washington Post*, 30 November 1988, p. A6.
74. Lev Mendelevich, Director of Policy Planning in the Soviet Foreign Ministry and reportedly involved in the evolution of this concept. Quoted in *Washington Post*, 30 November 1987, p. 6.
75. Quoted by Congressman Les Aspin, 'Conventional Forces in Europe: Unilateral Moves for Stability', *Bulletin of the Atomic Scientists*, December 1987, pp. 12–15. See also joint Soviet-Czechoslovak Communiqué in *FBIS*, 15 April 1987, p. F20.
76. Colonel General Chervov, Chief, Treaty and Legal Directorate of the USSR Armed Forces, *FBIS* – Soviet Union, 1 December 1988, pp. 103–4.
77. Gorbachev's speech to the United Nations, 7 December 1988. Reprinted in *New York Times*, 8 December 1988, p. 16. Air assault and river crossing units and equipment were also singled out for reduction. The reference to 'more defensive postures' may refer to the increased fire-power of Soviet operational maneuver groups that underwent considerable offensive reorganization in the 1980s. (See Note 51.)
78. Colonel A. A. Sidorenko, op. cit., p. vii.
79. Quoted in the *New York Times*, 7 March 1988, p. 1.
80. Ibid., p. 1. Students of Clausewitz may not be comforted by this.

Clausewitz argued that defense alone could not win wars. It had two phases; waiting for a blow and parrying it. Choosing the right time and place to unleash that 'flashing sword of vengence' was described as 'the greatest moment for the defender'.

81. Quoted in the *New York Times*, 7 March 1988, p. 1. This theme was also developed earlier in *Red Star* by Colonel General D. Volkogonov, 22 May 1987, translated in *FBIS*, 4 June 1987, pp. V4–V6.
82. 'A Most Reliable Defense – Above All', *Moscow News*, No. 8, 21 February 1988, p. 12.
83. Ibid., p. 12.
84. Asymmetrical reductions in Europe realistically mean actual reductions of men and divisions as well as redeployment for the Soviets. For NATO and the US in particular, reductions will (or should) mean redeployment only.
85. *New York Times*, 8 December 1985, p. 16A. Gorbachev first made this argument in his address to the Twenty-Seventh Party Congress in February 1986. 'The Character of Contemporary Weapons', in his words, 'does not permit any state to defend itself by military-technical means alone . . . .'

3 Soviet Incentives for Conventional Deterrence

'Show me a country without nationalist problems, and I will move there right away.'

Mikhail S. Gorbachev

Soviet leaders have compelling reasons for building a new strategic relationship with the United States. A post-nuclear world would be far more hospitable to the unique security requirements of the USSR.

Western concepts of nuclear deterrence are too often presented abstractly, outside the political context in which nuclear war would occur. The strategic debate has been dominated by the visible indicators of military power–delivery vehicles, nuclear warheads, throwweight, and accuracy for example. The credibility of nuclear deterrence must be viewed across a wider spectrum of variables if destabilizing trends in either force structure or strategy are to be avoided and the arms competition they foster checked. Technical capabilities and quantifiable threats must be linked more precisely with the full range of threats faced by each country. These include the geopolitical, economic, ethnographic, and even the historical factors that influence the calculus of Soviet strategic planning.

Soviet sensitivity to homeland defense is far more complex than is generally recognized in Western discussions of nuclear deterrence and war. Homeland defense requires more than a robust capability to guard Soviet borders and maintain territorial integrity. In Soviet eyes, a credible homeland defense must also:

(1) Maintain ethnic Russian domination of a multinational State.
(2) Maintain Communist Party control of both the ethnic Russian heartland and the strategically located, non-Russian Union Republics which make up the USSR.
(3) Maintain the current political élite's personal control of the Communist Party.
(4) Provide the military forces that give the Soviet Union superpower status.

The first two of these four interrelated security objectives are unique to the Soviet State. They are unique by virtue of the anachronistic style of Soviet Communism–a relic of nineteenth-century Western political thought that has fastened tenaciously onto the twentieth-century's last remaining empire. This empire was forged over several centuries by Russian Czars who successfully acquired power to take the offensive against waves of invaders who had repeatedly subjugated Russia. Centered in a vast geographic area which lacked natural frontiers or defensive barriers, the Czars both defended the State and satisfied personal ambitions for power by expanding Russian frontiers. The results of this expansion are seen today in the administrative structure of the USSR. Its 15 Union Republics are organized around the Soviet Union's dominant ethnic and cultural groupings–the nationalities as Soviet officials describe them (see Map 3.1).

Maintaining ethnic Russian control of the nationalities is one of the most serious strategic challenges to Soviet leaders in both war and peace. Yet geopolitical vulnerabilities to societal disruption and political fragmentation are among the least examined variables in the assessments of Soviet military power and risk-taking. Western strategic literature treats the Soviet Union as a unitary state, powerful in its military and political potential to threaten the United States and its allies. Little has been done to examine the multinational character of the Soviet state and its potential effect on the Soviet-American mutual deterrence relationship. Both official and popular American images of the Soviet Union rarely recognize the multinational and formally federal character of the Soviet State. The Communist Party, dominated by ethnic Russians, should not be confused with the more complex multinational state.

Ethnic Russians soon will comprise a minority of the Soviet population (see Table 3.1).[1] They are concentrated in the center of the USSR (The Russian Soviet Federated Socialist Republic or RSFSR, one of 15 Soviet Republics), and are buffered from neighboring countries by Union Republics populated predominantly by non-Russian ethnic groups (See Map 3.2). Most important, many of these ethnic minorities have long histories of political independence. How Soviet leaders have managed pressures for autonomy or independence by these groups during periods of crisis or national stress tells us a great deal about Soviet perceptions and sensitivity toward these points of vulnerability. World War I and the Bolshevik Revolution, for example, led to temporary independence for some

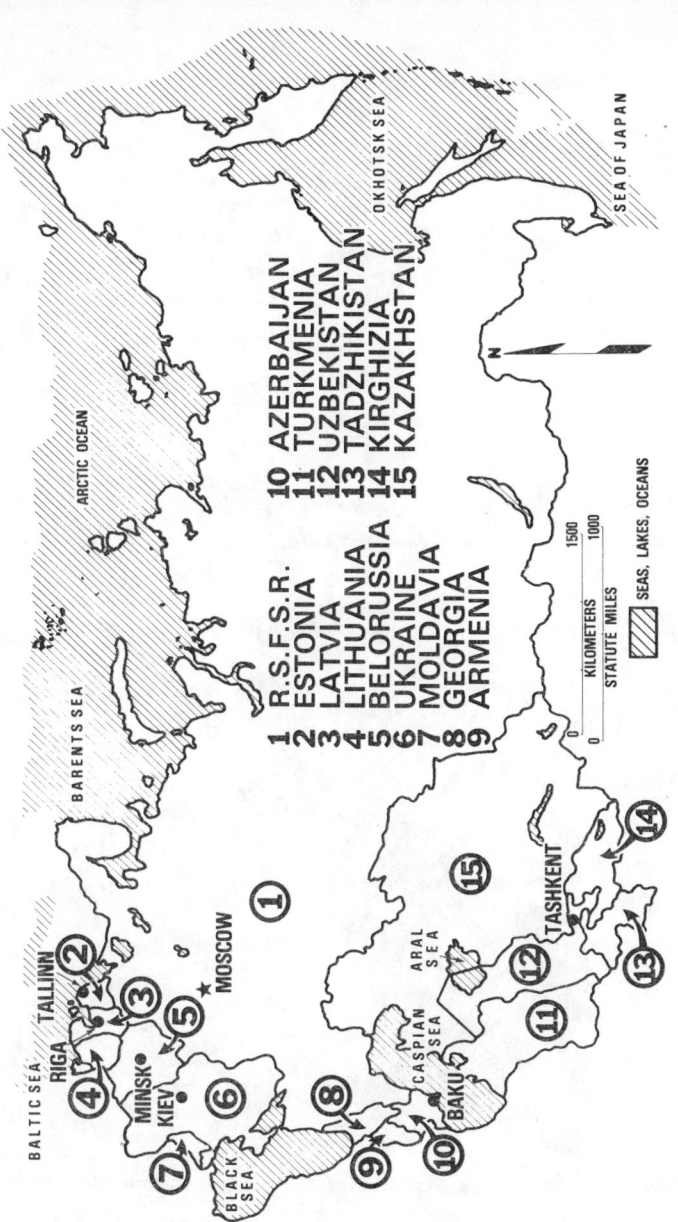

MAP 3.1 Union republics of the USSR

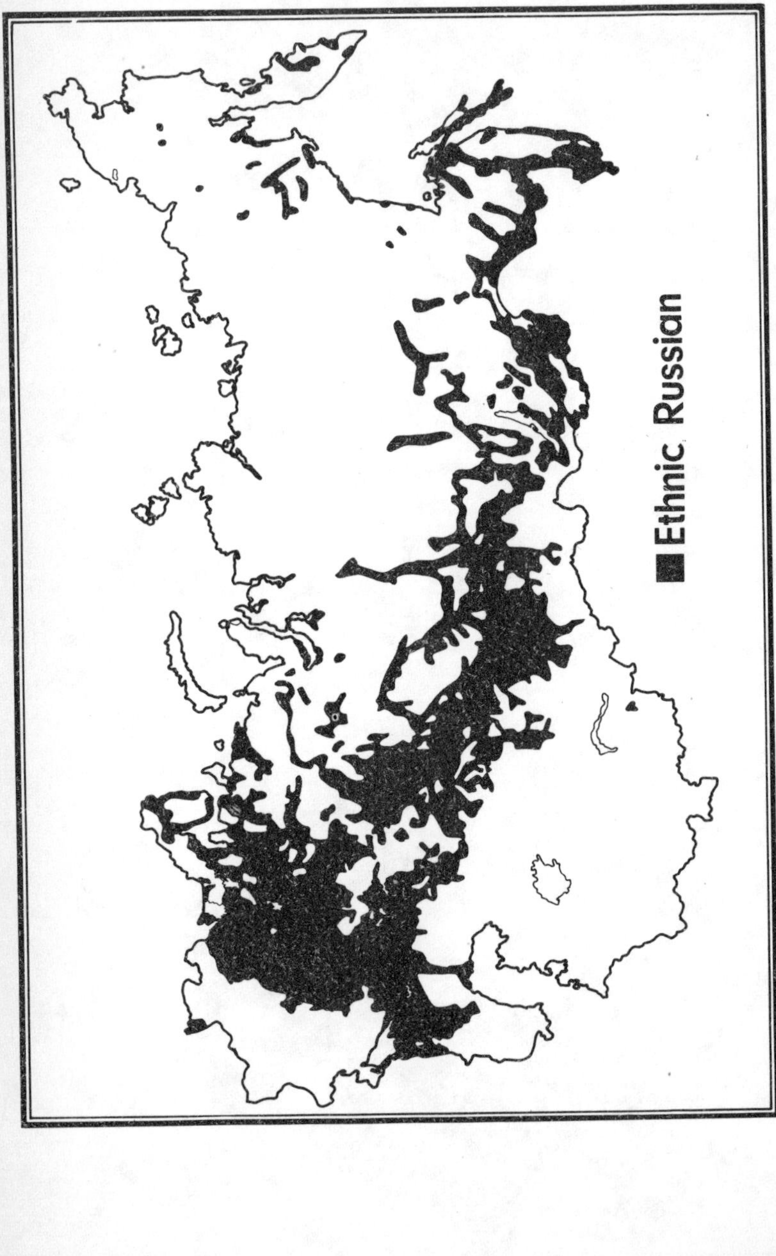

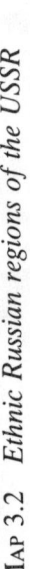

Map 3.2 *Ethnic Russian regions of the USSR*

TABLE 3.1 *Shifting population trends*

				Total
1985	Russian	143 585 000	51%	
	Non-Russian	136 212 000		279 797 000
2000	Russian	151 350 000	48%	
	Non-Russian	162 355 000		313 705 000

Source: U.S. Census Bureau

ethnic groups, which later had to be forcibly reintegrated by the Red Army. Similarly, during World War II Stalin relocated entire ethnic populations to the interior of the country for fear that they might collaborate with the Germans. Nor was this fear unwarranted. Many groups did defect in large numbers, taking up arms on the German side. As the German armies moved through the Ukraine and Belorussia, they were conquering regions that had been the most cruelly hit by forced collectivization of 1930–33, famine, and Stalin's russification policies. Had the Germans been capable of humane and moderate treatment of the Soviet nationalities in these areas, their occupation could have become a danger to the Soviet system even after the German retreat. One can only speculate as to what additional problems the Soviets would have encountered had Hitler in 1941 proclaimed the independence of the Ukraine, Belorussia, and the Baltic states. According to Adam Ulam, the continuous demands for a second front in Europe, even when the Germans could no longer win in the east were prompted by the urgent necessity of reconquering Soviet territories as soon as possible and before any form of anti-Soviet organization could take root.[2] As it was, pockets of anti-Soviet partisans in these areas resisted the Soviet Army for several years following the German surrender in 1945.[3]

More recently the resurgence of Islam in combination with increased ethnic nationalism on or near the Soviet border has increased the possibility that their own Islamic and minority populations in the areas bordering Iran and Afghanistan may in the future press for greater autonomy. Once set in motion, the pressures of nationalism could start several ethnic dominoes falling out of control.

Parallels can be drawn between the Soviet invasions of Afghanistan and Czechoslovakia. Soviet sensitivity to events in Iran and Afghanistan is undoubtedly heightened by the potential impact of political and religious ferment in these areas upon Soviet Islamic citizens in Central Asia. A similar situation existed in Czechoslovakia where reforms had an unsettling effect on autonomy-minded Ukrainian nationalists.[4] The Ukraine had developed close cultural and economic links with Czechoslovakia. This, in combination with a small Ukrainian population in Slovakia (the number would be greater if Stalin had not annexed Polish territory in 1939 and the Carpatho-Ukraine in 1945, thereby extending the Soviet border to Czechoslovakia and Hungary, facilitating the projection of military power into those countries, and minimizing future conflicts that might arise between Ukrainians and Eastern Europeans), resulted in Ukrainians being more exposed to the reformist and nationalistic ideas expressed in Czechoslovakia. This exposure, superimposed upon indigenous nationalism, resulted in a breakdown of the official Soviet monopoly of the means of public communication and political indoctrination. According to the 'Ukrainian hypothesis', for Soviet officials, no 'mental frontier' separated the Czechoslovak crisis from the Ukrainian problem.[5] The nationality problem played a dominant role in shaping the Soviet decision to invade and crush the 'Prague Spring'. According to this thesis:

> Czechoslovakia would have appeared in the mind's eye of the Soviet leadership as a union republic in which the 'bourgeois nationalists' were actually getting away with what 'they' were trying to do in the Ukraine . . . . The definition of and response to the Czechoslovak situation . . . would be considered from *this* perspective as a projection outward of a campaign underway already in the Ukraine and other national republics to combat local nationalism and anti-Russianism. The critical factor here would be the cognitive impact that Ukrainian dissent had presumably already made upon the Soviet leadership.[6]

A more recent study of Soviet ethnic problems also identified the Ukraine as the key to ethnic stability in the USSR. The Ukrainians, of all the non-Russians, have the absolute strength to pose a serious threat to the Soviet state on their own. Their regional hegemony based on political, economic, social, and cultural resources could make the Ukraine the USSR's Poland.[7]

The precise relationship between contemporary Soviet domestic and foreign policies cannot be stated without firsthand knowledge of Soviet decision-making. Whatever the linkage may prove to be, there is little doubt that Soviet domestic vulnerabilities are taken into account during times of crisis, and play a role in Soviet assessments of both their conventional and strategic force requirements. The nationalities issue is especially significant in assessing Soviet vulnerability to nuclear war.

STRATEGIC IMPLICATIONS OF THE NATIONALITIES

Nationalism in the Union Republics remains a problem for Soviet leaders much as it was for their Czarist predecessors. Marxist-Leninism has not produced a melting pot for proletarian internationalism even within the borders of the USSR. Under Stalin, the rhetoric of 'friendship of peoples' characterized the federal structure of the USSR, masking both his ability and willingness to deal harshly with troublesome and untrustworthy non-Russians in the Soviet borderlands. Khrushchev reopened the 'nationalities problem' by emphasizing the need to equalize rates of economic development and provide equal opportunities for all Soviet nationalities. His 'affirmative action' policies stressed building Communism and *merging* all Soviet nations into a higher community–'the Soviet People'. Under Brezhnev, less ambitious attitudes emerged in discussions of the new Soviet Constitution. For example, in a remarkably candid public confession published in 1977, Brezhnev admitted that the 'merging' of the Soviet nationalities had given way to 'rapprochement' and declared that 'we would be entering a dangerous path if we were to artificially force the objective process of the rapprochement of nations'. Instead, he foresaw a long-range process of 'nations drawing together'.[8]

Yuri Andropov displayed great sensitivity to the nationalities question during his brief reign. He reasserted the Leninist idea of a 'merger' of nationalities as the long-term goal, but emphasized economic integration and equality rather than ideology as the primary vehicle for national cohesion. National distinctions will exist longer than class distinctions. Moreover, Andropov saw potential strife when he warned that economic progress among the various nationalities would inevitably be accompanied by the growth of national self-awareness. Ethnic pride, he warned, should not degenerate into ethnic or regional arrogance. Economic progress and the migration of population required for labor mobility (and control) has made each republic more multinational. This means the Party and Government 'must carry forward lofty principles' to ensure harmonious and fraternal relations among ethnic groups.[9]

Mikhail Gorbachev was slow to address the nationalities question and this suggests that the issue was not at the forefront of his concerns. The problem had been secondary to his broader goals of economic reforms and progress. In his drive for economic efficiency, the Soviet leader has shown impatience with 'parasitic attitudes' of some republics.[10] This impatience could be seen in his sacking of Dinmukhamed Kunaev, the local Party chief and full Politburo member from Kazakhstan. Riots followed in the capital city, Alma Ata, after Kunaev was replaced by an ethnic Russian.

There is evidence that the riots were encouraged by local Party members who feared with good reason that the fall of their patron would cost them their positions.[11] Local resentment, however extensive, seems to have been effectively dissipated by the new leadership's ability to quickly get meat and vegetables in state stores. Previously, one-third of Alma Ata's food supply and 80 per cent of its housing had been siphoned off for the Party and State élite. An honest Russian who can show results may be preferable to a corrupt kinsman. Gorbachev has clearly stated his preference for economic efficiency even at the cost of local ethnic resentment at reforms which sweep local leadership away. Nevertheless, there are risks, and as the riots in Kazakhstan and elsewhere demonstrate, ethnic sensitivities can be easily manipulated. *Glasnost* or greater openness may lead to greater ethnic identification and assertiveness–a trend not welcomed by hardliners concerned with maintaining Russian control.[12] The volatility of *glasnost* and ethnic nationalism has become increasingly evident. Map 3.3 shows the ring of ethnic unrest in several Union Republics that has been rekindled since 1987. The most dramatic episode was in Armenia. The largest protest demonstration in Soviet history took to the streets in support of demands by ethnic Arme-

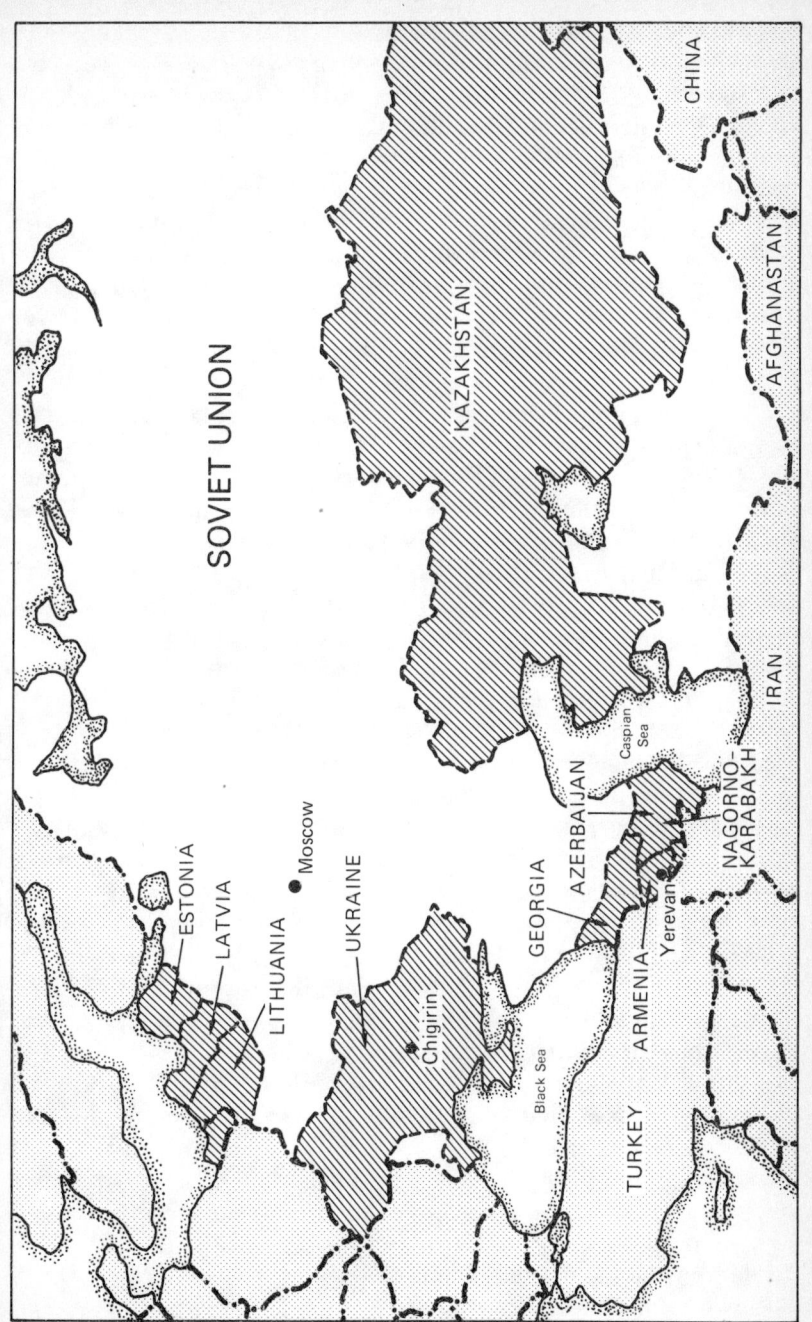

MAP 3.3 *Areas of Ethnic Unrest since Glasnost*

nians in the neighboring republic of Azerbaijan to incorporate the region of Nagorno-Karabakh into Armenia. These clashes were not overtly anti-Soviet (some Armenians carried photos of Gorbachev), but Soviet leaders nevertheless fear a domino effect. Azerbaijanis are predominantly Shiite Moslems, the sect most open to Islamic fundamentalism. The population of 5.5 million is matched by an equal number of Azerbaijanis in northern Iran. The proximity of Soviet Shiite Moslems to Iran, the source of anti-Soviet radio broadcasts into Central Asia, and an active Islamic underground make the authorities uneasy. A KGB security chief in Tadzhikistan, for example, reported that his republic was a 'fertile ground' for Islamic ideas and groups opposing the Soviet system.[13]

Elsewhere, nationalism took more immediate anti-Russian forms. The riots in Alma Ata were followed by protesting Crimean Tartars in Moscow who called on Gorbachev to restore the Crimean homeland from which they were deported by Stalin in the 1940s. Thousands of demonstrators also poured into the streets of capital cities in Latvia, Lithuania, and Estonia protesting the 1939 Nazi-Soviet Pact that permitted the Soviet takeover of the region.[14] The Estonian Central Committee went so far as to pass local measures to reduce the influx of Russian emigrants. New regulations require local firms to pay 16 000 rubles for every employee they bring from another republic, and an additional 16 000 for each family member.[15]

In Soviet Georgia, troops violently cleared the streets of Tbilisi where demonstrators were protesting Moscow's efforts to end Georgian control over the small autonomous republic of Abkhazia. Despite decades of Soviet rule, and, for most of the Soviet Union, another century under the czars, most Union Republics maintain a powerful sense of nationhood or a distinct cultural identity complete with ideas on where their borders should lie and on how they should be allowed to deal with their own minorities within those borders. There are many layers of ethnic conflict and nationality issues that *glasnost* has brought to the surface.

It is not surprising that Mikhail Gorbachev told a Party Central Committee meeting that '...at the present stage we must get down to some very substantive work on nationalities policy,... this is a most fundamental and most vital question facing our society. I think we ought to devote a Central Committee plenum to the problems of nationalities policy'.[16]

The irony he faces is that domestic reforms have required renewed de-Stalinization to reverse the structural legacy of his era. De-

Stalinization, in turn, opens his dark record of massive injustices against many non-Russian nationalities. Changes in economic and social conditions may, as Andropov feared, increase ethnic identification and resentment of assimilationist pressures from central authority. Increases in ethnic tensions seem more probable than wishful Soviet predictions of 'nations drawing together' unless Soviet leaders are skilful enough to avoid the tensions produced by ethnic Russian domination of political and economic institutions. Economic resources have been spread generously throughout the Soviet Union's non-Russian republics, but prosperity is no deterrent to separatism. History demonstrates that revolutions, including Lenin's, are made by educated, reasonably prosperous élites.[17]

From the Soviet perspective there are additional unsettling precedents in their foreign policy which inadvertently foster nationalism among their own minorities. Support for the Arabs after the 1967 war against Israel, for instance, was a significant factor in provoking a resurgence of Jewish nationalism and the desire for increased emigration. By the same token, success of Jews in establishing their right to emigrate (however limited) has stimulated a similar cause among Baltic Germans.[18] Confronted by a pattern of non-Russian self-assertiveness, assisted perhaps by the US human rights campaign, Soviet officials may well speculate that today's would-be emigrant could be tomorrow's separatist. Similarly, Moscow's support and encouragement for national liberation movements has also backfired to a certain extent. Many nationalist writings have pointed to the incongruity between Soviet foreign and domestic policies.[19] There is no small irony in the world's largest multinational state, or more accurately, empire-state, being the leading spokesman for national liberation movements.

None of these indicators proves that the disintegration of the Soviet Union is an immediate or even long-term prospect. The Soviet Government shows every indication of being able to deal with its internal problems. But how these problems would pose themselves during periods of crisis and societal disruption are an entirely different matter. War, and most particularly nuclear war and its aftermath, requires a unity of effort that some fear might be lacking even in the United States. The Soviet problem would be far more complex and uncertain. Wars of this century have not fostered optimism in the Russian soul. Beaten by the Japanese in 1905, by numerically inferior, but technically superior, Germans in World War I, by the Poles in 1920, by the Germans again in 1941–42, the

Soviets triumphed in 1943–45 because of a remarkable mobilization of patriotism. Could they do this again after a nuclear war or during a general war in Europe? Non-Russian minorities have little interest in the objectives of Soviet foreign policy. They detract from Soviet externally projectable power. The point is not the likelihood that they will rise up *against* the Soviet State. The point is they have little reason to rise up *with* the State. All previous efforts by Soviet leaders for the merging of regional loyalties with the State's political objectives have falsely assumed that the ideologically, class-based integration of the Soviet nationalities could occur in a state where a Russian national ethos more than ideology remains the dominant basis for political loyalty. Nowhere is this more apparent than in the structure of the Soviet armed forces.

The military aspects of Soviet integration (namely, russification) policy provide additional clues about doubts Soviet officials may harbor about the *crisis loyalty* of their nationalities. Major combat units of the army are dominated numerically and administratively by ethnic Russians. Less skilled minority recruits are more likely to be assigned to support roles. In the majority of instances, the latter are garrisoned at bases outside their homelands.[20] No nationality group has large concentrations of native troops stationed on its own soil. This was common practice even before the new Soviet Constitution formally dropped the Union Republics' right to possess their own armed forces. In short, the ethnic dispersal of the Soviet army often results in colonial-like occupation patterns in many areas where troops find themselves in unfamiliar social milieu, climate, and a culture sharply at variance with their own. In turn, they are sometimes regarded with disdain by the people whose territories they occupy–even within the USSR.[21] Decades of forcing non-Russian conscripts into labor units to fetch and carry for Russian soldiers has exacted its price.

Party-Government administrative control in non-Russian areas is also structured to check the emergence of autonomy-minded local bureaucrats. First or second Party secretaries are nearly always ethnic Russians in whose hands ultimate decision-making power resides, including the recruitment of local administrators.[22] Russian-dominated local bureaucracies have been accompanied by large influxes of Great Russians into urban areas where they have dominated the process of modernization and industrialization, and have benefited disproportionately from the higher living standards that result.[23] These patterns seem likely to increase ethnic tensions, especially in the immediate future, as the Soviets are forced to deal with a declining labor force in the RSFSR. Their choices include expanding industry in non-Russian Republics where labor is more plentiful, or bringing more minority labor into the predominately Great Russian RSFSR. Either option risks increasing domestic tensions in a nation that has always seen a close relationship between domestic stability and military power.

IMPLICATIONS FOR US STRATEGY

Conducting offensive operations while maintaining a stable home front places unique pressures on the Party and the General Staff. Surely any responsible leadership would harbor the gravest doubts as to the adequacy of Soviet strategic and conventional forces to underwrite the enormous wartime demands placed on them by Soviet doctrine. As Benjamin Lambeth has pointed out, 'Because ...obligations place open-ended demands on Soviet force availability, performance, and durability, the Soviet leaders can never feel so complacent about the adequacy of their preparedness efforts as to permit any prolonged resting on their strategic oars'.[24] Lambeth's observations are in sharp contrast with the often repeated belief that Soviet military preparedness goes far beyond legitimate defense requirements. If correct, Soviet notions of 'sufficiency' and 'homeland' defense are inevitably going to be considerably more ambitious than their American counterparts. 'Mother Russia' (The RSFSR) is surrounded by non-Russian Republics, which are bordered by subservient, but unreliable allies which are, in turn, surrounded by hostile neighbors and military alliances. These combined threats to Soviet security may do more to strengthen the credibility of US deterrence while undermining the Soviet appetite for risk taking than any variant of military hardware or technical capability. At best, nuclear weapons are an imperfect means of compensating for the geopolitical liabilities unique to the Soviet State. These liabilities place serious constraints on the use of Soviet strategic forces as tools that can be employed in preplanned ways to coerce concessions from an adversary, or that might tempt Soviet leaders to reckless and inflexible positions during crises.

Nevertheless, there is little doubt that the Soviet strategic build-up since the 1960s has contributed to major changes in US strategic doctrine and force structure. The size and characteristics of US strategic forces have been determined by the requirements for putting at risk specific Soviet target categories. What those targets should be are often the subject of rigorous debate. One reasonable objective is the erosion or elimination of Soviet confidence in military solutions to crisis. As Colin Gray has put it:

> One of the essential tasks of the American defense community is to help ensure that in moments of acute crisis the Soviet general staff cannot brief the Politburo with a plausible theory of victory.[25]

A Soviet decision to go to war requires much more than military confidence of the General Staff. The Soviet calculus requires political, social and economic confidence as well. This presents US strategists with a broad deterrence spectrum in threatening those interrelated values that will most credibly prevent Gray's 'victory' briefing from becoming plausible. What kinds of threats would have the most deterring effect on the Soviet leadership? The Soviet's nationalities problem is relevant to US strategy. The non-Russian populations are a political center of gravity for the cohesion and integrity of the State. They are also a center of gravity in any Western pre-war deterrence or wartime strategy aimed at disrupting the Soviet rear. There are parallels here with counterinsurgency. Insurgents depend on the population for their long-term success. A besieged government must also draw support from the same population if it is to survive and defeat an insurgency. Both insurgents and government have the same center of gravity. In both counterinsurgency and strategic nuclear war a common dilemma in formulating strategy is how to attack an enemy without threatening a center of gravity which is important to your own success. In the present case, the Baltic States, Ukrainians, Central Asians, Georgians, and other ethnic groups are not the enemy. Indeed, they are the potential vehicles for disrupting the Soviet rear. Attacking them directly would be as counterproductive as widespread and indiscriminate civilian casualties in counterinsurgency operations.

The military-economic center of gravity in the USSR is that section of the RSFSR from the Urals to its Western boundaries. Here is concentrated the largest percentage of ethnic Russians, ICBM fields, naval facilities, bomber bases, industry, communications, and transportation facilities. The threat of assured and concentrated retaliation in the RSFSR confronts Soviet leaders with the prospects of a radically altered domestic and international balance of power.

For the Soviets, recovery would be complicated by political problems they would confront in the presumably less damaged non-Russian Republics. Could the economically linked, but physically less damaged zones be counted on for recovery assistance as in the case of other localized disaster recovery efforts? Or would scarcity and chaos further stimulate the centripetal forces of nationalism and separatism? Many of the outlying Union Republics served their buffer functions well in World War II, absorbing the initial damage and destruction of the German Army. In a nuclear war, the reverse may be true. The central Great Russian zones (RSFSR) could receive immediate and highly concentrated levels of damage.

The evolution of US nuclear strategy toward flexibility, proportionality, and controlled responses has produced a force structure that is capable of some level of political discrimination. This does not mean that credible deterrence demands extensive threats to Soviet industry or Russian population centers. The destruction of essential choke-points in a highly interdependent economic system would effectively shut down industrial production, even if many plants and industrial centers survived. Soviet sensitivity to threats aimed at the industrial infrastructure which supports its superpower status, combined with its strategic perceptions that long wars require a stable political and economic base, suggest that a small, but survivable nuclear force may provide an effective, but thus far insufficiently explored basis for achieving stability, credible deterrence, and strategic arms reductions.

Ironically, Soviet vulnerability may also contribute to their reluctance to endorse American limited nuclear war fighting strategies. Vulnerability reinforces the value of deterrence. If deterrence fails, limited nuclear attacks may be viewed in Soviet eyes as already doing or about to do our worst. In war, attention is focused on overall vulnerability, not the actual scope of the attack at a given moment. Limited nuclear dueling, therefore, is an unlikely Soviet response.

The Soviets, understandably, do not openly discuss or link the nationalities question to strategic vulnerability in war. Their actions, however, indicate that the leadership harbors serious misgivings about the crisis loyalty of many Soviet minorities. If so, these doubts contribute to Soviet self-deterrence and a preference for low risk taking in crises involving the threat of confrontation with US strategic forces.[26] It is not in the American interest to shine too bright a light on Soviet nationality problems. There would be a significant danger and probable Soviet backlash if American officials initiated widespread discussion of Soviet ethnic vulnerabilities with no accompanying restraints in the form of offensive arms control and general improvements in Soviet-American relations. Their heightened perception of US hostility could easily prompt countermeasures and an escalation of the arms race that would only lock the United States into futile action-reaction spirals that do little to increase security.

Recognition of Soviet weaknesses and their impact on Soviet homeland defense serves to strengthen confidence in and the credibility of existing US strategic doctrine and force structure. If and when that force structure declines as the result of arms control agreements, greater efforts will be required to maintain deterrence and economy of force. This will require a more precise definition of the Soviet centers of gravity. Linking deterrence strategies and Soviet multinationalism is one possible approach under the administration's competitive strategies initiative.[27]

The Soviets are fully aware of their enduring political liabilities. These liabilities provide a considerable Soviet incentive for superpower stability (peaceful co-existence). If and when Soviet leadership shows a preference for conflict, the preference will most likely flow from a posture of conventional superiority. Their most useful military tools are conventional forces where Soviet capabilities are less encumbered by the self-deterring pressures that are found in Soviet nuclear risk-taking behavior.

The USSR's most enduring weakness is its political and economic structure. The Gorbachev domestic agenda may signal a new, more co-operative phase in Soviet-American relations and ultimately, a stronger, more competitive Soviet industrial base. No one can say whether a rehabilitated Soviet socio-economic system would spawn a more assertive foreign policy or a status quo mentality anxious to preserve the benefits of reduced tensions abroad and higher living standards at home. In a world of uncertainties, American security requires the patient, but long-term maintenance of credible military forces and aggressive political efforts to improve Soviet-American relations on all fronts. Successes in a co-ordinated approach to strategic, theater, and conventional arms control are an essential ingredient of any post-nuclear relationship between the superpowers. The incentives are present. The question of trust is more divisive, and may pose a serious threat to any American president who attempts to travel too far on the arms control path. This challenge is discussed in Chapter 4.

NOTES

1. Some experts had predicted that the 1979 census would show ethnic Russians to be a minority. The published Soviet statistics showed ethnic Russians as 52.4 per cent of the population. Murray Feshbach of Georgetown University's Center for Population Research predicts that figure will fall to 48 per cent by the year 2000.
2. Adam Ulam, *Expansion and Coexistence: Soviet Foreign Policy 1917–73* (New York: Praeger, second edn, 1974), pp. 326–7.
3. Reported in the *New York Times*, 19 April 1946, p. 19, 15 May 1949, p. 1, 26 July 1949, p. 9, and 1 May 1950, p. 10. A detailed account can be found in Frederic N. Smith's, 'The War in Lithuania and the Ukraine Against Soviet Power', in *Combat on Communist Territory*, Charles Moser (ed.), (Regnery Gateway, 1985).
4. Grey Hodnett and Peter Potichnyi, *The Ukraine and The Czechoslovak Crisis* (Canberra, Australia: Australian National University, 1970).
5. *Ibid.*, pp. 121–5.
6. *Ibid.*, pp. 124–5.
7. Alexander J. Motyl, *Will the Non-Russians Rebel?* (Ithaca, New York: Cornell University Press, 1987), pp.xi, 50.
8. Quoted in A. Shtromas, 'The Legal Position of Soviet Nationalities and their Territorial Units According to the 1977 Constitution of the USSR', *Russian Review*, July 1978, p. 272.
9. Extensive treatment of the nationalities was given during his speech before the Supreme Soviet celebrating the sixtieth anniversary of the USSR. Reprinted in *Reprints from the Soviet Press*, Vol. XXXVI, No. 1, 15 January 1983, pp. 8–18.
10. For an assessment of Gorbachev's initial policies toward the nationalities see the analysis of his speech before the Twenty-Seventh Party Congress and Roman Solchanyk, 'Does Gorbachev Have a Nationalities Policy?', *Radio Free Europe-Radio Liberty RL 112/86*, 7 March 1986.
11. *Washington Post*, 22 February 1987, p. A1.
12. This experience may explain why the Ukrainian Party Boss, Vladimir Shcherbitsky, survives as the last Politburo member of the Brezhnev era in spite of intense criticism of local corruption and pressure from the new General Secretary.
13. *New York Times*, 6 March 1988, p. 3f.
14. *Washington Post*, 9 and 27 August 1987, p. A1.
15. *Washington Post*, 15 March 1988, p. A35. This amounts to $100 000 per family of four.
16. 18 February speech printed in *FBIS*, 19 February 1988, p. 49.

17. See, for example, the literature on 'relative deprivation'. Ted Robert Gurr, *Why Men Rebel* (Princeton, N.J.: Princeton University Press, 1970), Chapters 2 and 3; James C. Davies (ed.), *When Men Revolt and Why* (New York: Macmillan, 1971), Chapters 2 and 3; and Crane Brinton, *The Anatomy of Revolution* (New York: Harper, 1938).
18. Julian Birch, 'The Persistence of Nationalism in the USSR', *Journal of Social and Political Affairs*, January 1976, p. 75.
19. *Ibid.*, p. 72. It is also ironic that at a time of unrest in many Union Republics, the Soviets were strongly endorsing similar activities in Gaza and the West Bank. See stories in *Pravda* reprinted in *Current Digest of the Foreign Press*, Vol. XL, No. 5, 2 March 1988, pp. 14–15 and *FBIS*, 2 March 1988, p. 20.
20. Jeremy Azrael, *Emergent Nationality Problems in the USSR*, R-2172-AF (Santa Monica, California: Rand, September 1977), pp. 16–22, and Sig Mickelson, 'USSR Muslim Population Explosion Poses Possible Threat to Soviet Military', *Military Review*, November 1978, p. 39. See also Susan Curran and Dmitry Ponomareff, *Managing the Ethnic Factor in The Russian and Soviet Armed Forces* (Santa Monica, Ca.: RAND, 1982).
21. Mickelson, op.cit., p. 39.
22. John H. Miller, 'Cadres Policy in Nationality Areas: Recruitment of CPSU First and Second Secretaries in Non-Russian Republics of the USSR', *Soviet Studies*, January 1977, pp. 8, 12, 18.
23. Robert Lewis and Richard Rowland, 'East is West and West is East ... Population Redistribution in the Soviet Union and its Impact on Soviet Control', *International Migration Review*, Spring 1977, pp. 6, 11.
24. Benjamin S. Lambeth, 'The Political Potential of Soviet Equivalence', *International Security,* Fall 1979, p. 37.
25. Colin Gray, 'Nuclear Strategy: A Case for a Theory of Victory', *International Security*, Vol. 4, No. 1, Summer 1979, p. 56.
26. There is nothing in the public record that shows the Soviets have ever placed their strategic nuclear forces on alert during a crisis.
27. 'Competitive Strategies' (see Chapter 5) emerged in Secretary of Defense Caspar Weinberger's FY 87 *Annual Report to Congress*. Its objective is to align enduring US strengths against enduring Soviet weaknesses. As a fiscal strategy it seeks to offset deficit forced budget constraints with more efficient use of resources.

4 Arms Control: Can We Make This Trip With the Russians?

Dramatic progress in arms control during the final years of the Reagan Presidency was all the more remarkable when compared with his first term in the White House. The commitments to arms control undertaken by three previous administrations were overshadowed by ambitious defense modernization programs and tough anti-Soviet rhetoric.

By 1985 the threat of nuclear war, costly arms competition, and domestic political pressures forced arms control back to a central position on the foreign policy agenda. Secretary of Defense Weinberger's *Annual Report to Congress* for Fiscal Year 1986 identified arms control as one of the 'four pillars of U.S. defense policy'. Certainly it is the primary pillar of the transition to a post-nuclear era in Soviet-American relations.

Yet, the Secretary of Defense and his subordinates did everything in their power to topple the arms control pillar. The campaign to break out of SALT II Treaty constraints and to broaden the administration's interpretation of the ABM Treaty, opening legal avenues for testing strategic defense components in space, were orchestrated from the Pentagon. The office of the Secretary of Defense (in sharp contrast to the Joint-Chiefs-of-Staff) was viewed in Washington as the major obstacle in the interagency development of arms control proposals.

The major weapon in the anti-arms control campaign was Soviet non-compliance with existing arms control agreements. The Soviets, it was argued, cannot be trusted, have cheated in the past, and will do so in the future. Arms control is a one-way street which leads only to Soviet advantages while lulling Americans into a false sense of complacency.

This chapter examines the extensive debate over arms control compliance. Is arms control a safe path to security, and can it remain a major component of Soviet-American relations? How accurate are the Reagan administration's charges of Soviet non-compliance?

Investigation of these questions focuses on the three agreements that have played the most important regulatory function in the Soviet-American strategic relationship–SALT I, SALT II, and the ABM Treaty. The Soviet compliance record is measured against three competing images: (1) deliberate cheating; (2) bureaucratic structure and process; and (3) conflicting interpretations of treaty obligations. Which of these images one accepts, quite clearly affects the desirability of entering into agreements with the Soviets in the future.

The data is drawn primarily from the congressionally mandated White House reports on Soviet non-compliance and the political debate surrounding those reports. Public Law 99–145 (Fiscal Year 1984 Arms Control and Disarmament Act) requires the President to submit classified and unclassified reports by 1 December of each year. While a complete evaluation of Soviet non-compliance is impossible when limited to unclassified data, even those with access to both reports have not always been able to reach a consensus on the scope of Soviet violations. Conflicting interpretations often divide members of Congress and the arms control bureaucracies within the executive branch despite their access to classified data and analysis.

This should not be surprising. Non-compliance is a conclusion drawn from the analyses of complex monitoring activities by the intelligence community and comparisons of Soviet behavior with treaty obligations. Intelligence data may be ambiguous *relative* to the legal complexities of treaty language. Intelligence ambiguity and legal complexity are hardly the characteristics of a process that cuts cleanly through rigidly held political/ideological views of the Soviet Union or arms control. Perceptions that the US is letting its guard down through arms control negotiations are certain to push the non-compliance debate to the front of the arms control agenda.

This assessment is designed to cut through the non-compliance debate, isolating what may be known from the often misleading political debate and selective intelligence leaks from bureaucratic insurgents bent on scoring political points at strategic times, for example, negotiating compromises or ratification debates. The purpose is neither to indict nor apologize for Soviet practices. The real questions are: is strategic arms control a viable process through which US security interests can be served?; can existing or more intrusive means of verification result in a more quiescent opposition to future Soviet-American arms control agreements?

The compliance debate has become a central part of Soviet-American relations for two reasons. First, the Reagan administration successfully linked future treaties to Soviet acceptance of extensive on-site inspection. Second, the administration built a case for Soviet non-compliance and has adapted it to the principle of 'proportionate response'. Proportionate response is based on a principle of international law that recognizes the right of one party to take action against the other in response to a material breach of an agreement. The administration has argued that since the Soviet Union has breached several existing arms control agreements including its political commitment to honor the unratified SALT II Treaty, the United States has the right to abandon its own pledge not to undercut SALT II and to take compensatory steps to protect its security. In practice this led to modest increases in the numbers of nuclear weapons above the SALT II weapons ceilings and a new, more permissive interpretation of the ABM Treaty, combined with an open debate on the virtues of abrogating that treaty in favor of extensive testing and deployment of space-based ballistic missile defenses. These radical departures leave the new administration with a political and legal responsibility to establish beyond a reasonable doubt that departures from existing arms control agreements are based on solid evidence of Soviet non-compliance. The Reagan administration position is captured in image one:

Image One: *The Soviets engage in a well-orchestrated strategy of non-compliance to gain unilateral military advantage over the United States.*

Testing this thesis requires careful examination of specific US allegations, a comparison of non-compliant with compliant behavior, and an assessment of comparative military gains and losses that may result from decisions to comply or push beyond legitimate treaty boundaries. Military gain can be assessed against three specific criteria:

(1) Does it contribute to a deployed military capability or alter the strategic balance?
(2) Does it improve a break-out capability by shortening deployment schedules?
(3) Does it undermine the predictability of Soviet strategic modernization programs?

Image Two: *Soviet Bureaucratic Politics and Standard Operational Procedures (SOPs) Create an Environment of Resistance, Subjective Interpretation, and Imperfect Oversight of Arms Control Obligations.*

Image Three: *Charges of Soviet Non-compliance Stem from Genuine Differences of Interpretation of Treaty Language and Obligations.*

The strength and validity of image one, on which the Reagan administration has based its case for Soviet non-compliance, can be tested against images two and three. Both tend to see a more compliant pattern of Soviet behavior. For this reason, as well as their close functional relationships, images two and three are examined together.

Image two incorporates bureaucratic politics and the standard operational procedures or organizational routines followed by the Communist Party, the Soviet military, and the strategic forces research and development infrastructure during the development, testing, and deployment phases of strategic weapons. Within many of these organizations and their sub-bureaucracies, the saliency of arms control concerns is almost certainly lower than are the broad priorities of Soviet defense policy.

The tone from the top can also set the overall standards for officials responsible for ensuring the compliance of their subordinate organizations. Leonid Brezhnev established a loose constructionist standard in his statement to Richard Nixon during the 1972 Moscow Summit when he indicated that everything not expressly prohibited by SALT I was permitted.[1] Unlike their American counterparts, Soviet bureaucracies lack the highly developed legal sieve through which their decisions on what may be permitted must pass. The process of legal counsel available to the US State Department, Arms Control and Disarmament Agency, Department of Defense agencies, and congressional oversight committees is far more centralized in the Soviet Union; and it may often be more restricted by Soviet walls of secrecy and compartmentalized systems of information. It is well established that Soviet lawyers, most of whom are attached to the Treaty and Legal Administration of the Foreign Ministry, serve the post facto role of legitimizing party decisions, and less the watch-dog function of their American counterparts. Such a system is prone to exploit the gray areas, and its actions may result in de facto violations, but not in the sense that they are coherent policies or pre-planned Politburo decisions to gain military advantage. Actions may be the result of subjective and unco-ordinated bureaucratic players.

It is also worth noting that Soviet bureaucratic routines may have gone through periods ranging from intense, competitive supervision to neglect during the last years of Brezhnev's life and the long period of maneuvering for political succession that lasted until Mikhail Gorbachev became General Secretary in 1985. Raising arms control compliance issues in the face of a highly competitive party structure during the early, most combative period of the Reagan administration could not have seemed wise or proper to most self-interested Soviet bureaucrats.

Image three is far less complex and assumes that considerable room for disagreement can develop from nearly any contractual agreement. To use a domestic analogy, tens of thousands of American lawyers earn comfortable livings from contractual disputes between parties who discover that clear-cut agreements grow ambiguous in practice. Soviet writings available in the West have not attempted to share arms control compliance guidelines that may be available to responsible bureaucrats. Soviet general attitudes toward monitoring have been discussed, however, and their basic principles vary considerably from those found in American discussions. For American negotiators, monitoring Soviet activities and verifying compliance is the vital substitute for mutual trust. Effective verification distinguishes the arms control process from the more utopian calls for world disarmament.

The Soviets themselves are concerned with verification, but they approach the problems with the traditional Russian passion for maintaining secrecy from the prying eyes of a hostile world. The 'basic principles' spelled out by the Soviets argue that the means of verification must be 'limited', 'in no way prejudice the sovereign rights of states', should not be built upon the principle of 'total mistrust', and should take into account the deterrent effect of the risk of detection by modern, sophisticated intelligence capabilities.[2]

To Soviet élites, the mutually agreed upon national technical means of monitoring Soviet territory and exchanges of previously classified data during SALT I and II were intrusive, if not revolutionary. One Soviet official is said to have remarked of these provisions, 'There goes 400 years of Russian history'.[3] It would be no surprise, then, to discover that decision-makers were more faithful to tradition than to the demands for sharing military information.

Determining when the political decisions were made that set disputed programs in motion is critical to this analysis. Certainly the decision dates for two major issues–the SS-25 missile and the Krasnoyarsk Radar–preceded the signing of SALT II.[4] Both parties attempted with great success to protect strategic modernization programs from arms control restrictions. How Soviet bureaucracies responded after 1979 is not clear. This period was the last chance to align modernization programs with treaty obligations.[5] This process, whether formal or informal, took place in the midst of both a growing succession struggle in the Communist Party of the Soviet Union (CPSU) and deteriorating Soviet-American relations. During this period, the CPSU's Politburo struggled to sustain the pretense that a faltering trio–Brezhnev, Yuri Andropov, and Konstantin Chernenko–were functioning national leaders. This struggle is difficult to reconcile with pre-planned cheating and deception directed from the top.

It was clear even before the Reagan victory in 1980 that SALT II was not going to be ratified. The period between 1980 and the spring of 1982–when both sides pledged not to undercut the provision of the SALT II Treaty–may have been a decisive phase inside the Soviet Government for arms control compliance issues. Strict constructionist positions that went beyond 'reasonable' Soviet interpretations of treaty obligations would have found a poor reception in the midst of competitive supervision produced by the struggle for succession to the Party leadership.

The transformation of treaty language into actions by rival Soviet decision-makers through self-interested bureaucracies during a period of growing US hostility certainly played a role in Soviet compliance behavior. The precise mechanics of Soviet decision-making are unknown in the West, but some general propositions can be explored on the basis of what is known about broader issues in the nuclear weapons decision-making process.

While there is widespread agreement that the Politburo is the center of political decision-making in the USSR, the complex machinations that take place beneath that body are clouded in secrecy. For example, the agenda for the Politburo is prepared by the Party Secretariat, an administrative body that normally includes at least three Politburo members. Does this overlapping membership mean that many issues have been effectively resolved and decisions drafted in advance of Politburo meetings? Does the Secretariat represent just one of many potential sources of influence? Are the overlapping members divided among themselves over basic issues? There are no Soviet sources that provide clear answers to these questions. Depending on the period under examination and the strength of the Party General Secretary, the answers could be yes or no.

A second organization, the Defense Council, also has a membership drawn from the Politburo. It is the highest Party-State body specializing in defense issues. From fragmentary evidence in Soviet sources this joint political-military committee chaired by the Party General Secretary manages national security issues on a day-to-day basis and shapes issues, including arms control, for wider politburo discussion.[6] The party leadership establishes guidelines for the military but, like our own National Security Council, neither the Defense Council nor the Politburo seem well suited to micromanage the technical details of defense or arms control.[7] Prior to Gorbachev, the Soviets never equated political control of the military with civilian management. Power was maintained primarily through control of personnel and resource allocation.

Khrushchev reported that he regularly saw blueprints of new aircraft, and that he initiated proposals for greater ICBM survivability, including details of silo construction.[8] Similarly, Brezhnev played a major role in ICBM development through direct contacts with design bureau chiefs.[9] It is difficult to imagine, however, that the Soviet leadership could sustain this level of intervention and discretion to the degree that their decisions could remain unaffected by the powers of bureaucratic competition that force even authoritarian and highly centralized ships-of-state to veer off course. For example, the Military-Industrial Commission, a State organization controlled by the Presidium of the USSR Council of Ministers, appears to have major responsibilities in weapons research and development and in controlling the system of scientific and technical information serving the defense sector. As a clearing house for technical information, the commission could play an important role in the internal arms control compliance debate. It is also worth noting that like the Party Secretariat and the Defense Council, the Council of Ministers includes Politburo members. The Chairman of the Council of Ministers or Premier is always a member of the Politburo, but not since Khrushchev has there been a General Secretary who also held this post. The overlapping, but not necessarily identical membership was intended to strengthen Politburo supervision and control. In practice, it also gives rival Party members access to powerful bureaucratic resources that can be mobilized in times of domestic or foreign policy crises.

The State Ministry of Defense has the primary responsibilities for the design requirements of weapons and military equipment produced by the various production ministries. The Soviet General Staff is subordinated directly to the Minister of Defense for the co-ordination of all the various service programs and activities. All evidence indicates that the General Staff of the armed forces is an extremely important player in the arms control decision-making process. Its members are the major source of information for the Defense Council and represent a mediating body between top political leadership and the military. The General Staff has played an active, if not central role in arms control negotiations through their participation with the Defense Council, Politburo, and direct representation on the SALT and START delegations.[10] The Ministry of Foreign Affairs has administrative responsibilities for the actual conduct of arms control negotiations, thus creating yet another layer of bureaucracy with possible supervisory compliance responsibilities. Treaty compliance debates may pass through all or only selected participants in these and other complex Soviet organizations.

We do not know the precise mechanics of Soviet arms control compliance. What is clear, however, is that rival factions in the Party have access to complex bureaucratic machinery managed by self-interested organizations. Competitive supervision of defense issues generally and arms control in particular cannot be documented. Absence of competition, however, would be an aberration of bureaucratic behavior within any political system–especially one undergoing an internal struggle for leadership that lasted from Brezhnev's rapid deterioration in 1975, through the brief Andropov and Chernenko periods, to late 1985 when Gorbachev appeared to gain control of the Party leadership.[11] During this period, frail Party Secretaries could not always count on the Central Committee to supervize their programs. Central Committee departments were as much a part of the succession struggles as the Politburo itself. The Ministry of Defense, the only major government ministry without a corresponding Central Committee department to oversee operations and administration, gained greater autonomy in the form of barriers to outside attempts by a fragmented Party leadership to micromanage their activities.[12]

The Defense Council may also have been less effective in its supervision of the Ministry of Defense and the armed forces. A fully engaged, unified Party management style would (and presumably does) have difficulty managing the technical details and standard operational procedures called for in Soviet arms control obligations. As Stephen Meyer has observed, there is a difference between 'decision authority' and 'decision influence'. No one questions the Party's authority to select options and have the final word in the decision-making process. The military and arms production infrastructure are subordinate to the Party, but their 'decision authority' in the form of their mastery of technical details and procedures is, nevertheless, powerful through their control of highly compartmented collection, processing, and structuring of data, policy alternatives, and operating procedures.

Steadily growing power in the hands of the military-technical community has been a visible trend because of the increasing technological content of defense and arms control policies, the military monopoly of analytic capabilities, and a lack of relevant politico-military experience or close personal military contacts among the emerging generation of Soviet political leaders. As Meyer concludes, technology and the military-industrial complex it has produced seem to have accomplished what nearly 70 years of Soviet Party-military relations have attempted to prevent–'the movement of the professional military into the heart of the defense decision-making structure'.[13]

There is nothing inevitable or irreversible in this trend. Soviet Party leaders have always been able and willing to micromanage high priority problems and projects while leaving lower priority issues to the relevant bureaucracies to manage in routine ways. As Charles H. Fairbanks concluded in a recent study,

> Soviet bureaucracy has undergone a succession of crises–the Revolution, the Civil War, the collectivization of agriculture, the terror, the German invasion–that habituated it to a continuous crisis atmosphere and required strict concentration on the most urgent problems ... Lack of resources and expertise made the rulers turn to solving problems by the concentration of will and enthusiasm on them.[14]

It is unlikely that arms control compliance issues received intense concentration or enthusiasm during the internal struggle for succession that was accompanied by a precipitous decline in Soviet-American relations during the final year of the Carter administration and the early years of Ronald Reagan. Arms control compliance could also have been affected by Soviet tendencies toward bureaucratic circumvention of central authority. Party-State directives and regulations are routinely rigid and cumbersome to an extent that adherence is often weakened. Bureaucrats who want to succeed must go outside the regulations to perform effectively. These practices are well known in the Soviet industrial sector.[15] Less is known about defense projects, but one can easily imagine short cuts to keep weapons testing and production programs on schedule and within budget and performance criteria. Bureaucratic survival tactics could explain several of the 'marginal' non-compliance issues described here.

At the very least, the growing breadth and depth of the military-industrial structure of the Soviet Union provides greater opportunity for disagreements, bureaucratic conflicts, and organizational routines that subvert compliance with complex arms control obligations. A paradox of bureaucracy is that policy can often be managed from below. By their actions or inactions, subordinates are capable of blocking policies, ignoring or disseminating reports, concealing activities or magnifying the importance of data. One of the lessons to be learned from bureaucratic politics on both sides is that negotiators should not resolve arms control deadlocks by falling back on ambiguous language that may later be exploited by political or bureaucratic factions.

The Charges

In January 1984, the administration sent to Congress the first in a series of mandated reports that concluded that there had been an 'expanding pattern of Soviet violations' of arms control agreements. The administration has submitted three follow-on non-compliance reports (in February and December 1985 and in March 1987) and the Arms Control and Disarmament Agency has published two additional reports of its own (in October 1984 and February 1986).[16]

The number of violations and the language used to characterize them have gone through considerable change (see Table 4.1). These changes suggest interagency conflict over the strength of new evidence and some new interpretations of old data. The President's reports to Congress conclude that there is 'a pattern of Soviet non-compliance'. The body of these reports contains ambiguities that may not support such a conclusion. For example, their language and categories make it nearly impossible to aggregate data in any 'clear pattern'. The reports discuss 'issues', 'violations', 'probable violations', 'likely violations', 'potential violations', and two questionable 'cases' where the Soviets were found to be 'in compliance'. These categories are further divided into classified and unclassified 'issues'.

TABLE 4.1 *President's Reports to Congress on Soviet Non-compliance with SALT I, SALT II, and the ABM Treaty*

Issues	Treaty	Jan 1984 report	Feb 1985 report	Dec 1985 report	Mar 1987 report
1. Impeding verification by encryption of missile test telemetry	SALT II	violation	violation	violation	SALT I and SALT II deleted from 1987 Report. U.S. no longer bound by these limits
2. SS-25 as a violation of the one new type ICBM provision	SALT II	probable violation	violation	violation	
3. Ban against SS-16 testing, production, or deployment	SALT II	probable violation	probable violation	probable removal of SS-16 equipment	
4. Strategic nuclear delivery vehicles exceed the 2504 cap	SALT II	–	–	violation	
5. Impeding verification by concealing the association between an ICBM and its launcher	SALT II	–	–	violation	
6. Use of dismantled SS-7 sites for support of SS-25	SALT I	–	no violation	violation	

7. Reconfiguration of one yankee SSBN for use as a cruise missile carrier	SALT I	–	no violation	–	
8. Krasnoyarsk radar	ABM	almost certainly a violation	violation	violation	violation
9. Development of mobile ABM system or components	AMB	–	potential violation	potential violation	potential violation
10. Concurrent testing of ABM and SAM components	ABM	–	probably has violated	probably has violated	probably has violated
11. Aggregate ABM activities provide base for territorial defense	ABM	–	may be preparing	may be preparing	may be preparing
12. Tested a SAM system or component as an ABM	AMB	–	–	insufficient to assess	insufficient to assess
13. Rapid reload of ABM launchers	ABM	–	–	ambiguous situation	ambiguous situation

The specific discussions of 'issues' and 'violations' are equally confusing. The December 1985 report, for example, lists six issues (discussed below) of concern about the ABM Treaty. Issues 1–5 end with the finding that the 'USSR may be preparing an ABM defense of its national territory'. Issue six is titled 'ABM Territorial Defense' and simply repeats the closing lines and several arguments of the previous five issues as a separate, sixth issue. Adding to the confusion is the fact that the case for possible Soviet non-compliance with the prohibited development of a territorial ABM defense is based on issues 1–5. Only issue one, the Krasnoyarsk Radar, is listed as 'a violation'. The remaining four issues are labeled respectively, 'potential violation', 'probably has violated', 'insufficient to assess', and 'ambiguous situation'. The March 1987 non-compliance report centers on these ABM Treaty issues. SALT II was deleted from the report as part of the policy ending US adherence to the SALT limitations.

The March 1987 report also introduced a new element in the efforts to assert that the Soviets were preparing a nationwide missile defense system. Three new Soviet radars were reportedly under construction on the Western borders of the Soviet Union. The Defense Department contended that these new radars were redundant with existing radars and appeared to be part of a missile tracking capability for an ABM system. The State Department, Arms Control and Disarmament Agency, and the CIA did not fully support this interpretation, pointing out that the radars seemed to be replacements for radars deployed in the 1960s. One radar could be intended to provide warning of attack by US Pershing-II missiles deployed in the Federal Republic of Germany. Whatever their purpose, these radars are among the first legally deployed systems that defense officials have attempted to associate with a major treaty violation. Skeptics believe that the issue may have more to do with administration efforts to break out of ABM constraints on SDI testing than with actual Soviet behavior.[17]

These examples of the administration's non-compliance reporting to Congress, and the interagency conflicts responsible for producing the often vague language with which they are written illustrate why the debate has often been confusing and frustrating to those who seek corrective actions. Confusing data and multiple categories of non-compliance have been amplified inaccurately by members of Congress and the news media. Cautious or inconclusive language was often collapsed into a single measure of Soviet duplicity. Analysts

110   *Deterrence and Defense in a Post-Nuclear World*

look in vain for any two sources that have cited the same number of Soviet violations. Part of the problem has been a failure to examine carefully what Soviet military incentives might be in a deliberate cheating scenario, and how those same objectives may be affected by uncontested Soviet compliance with the major provisions of SALT I, SALT II, and the ABM Treaty.

The Evidence

Table 4.2 summarizes the six issues on which the administration has concluded violations have occurred. Competing images and the potential military significance of Soviet non-compliance will be limited to these issues:

TABLE 4.2   *Competing Images of Soviet Non-Compliance*

Issues*	Image – 1 military significance of deliberate cheating			Image – 2 bureaucratic culture and SOPs	Image – 3 conflicting interpretations of treaty obligations
	strategic balance/ stability	breakout potential	undermines predictability		
1. encryption of test data	● no significance at SALT II levels	● marginal depending on other monitoring capabilities	yes	● tradition of secrecy ● complex encryption capabilities standard on all test missiles ● compliance over sight difficult in complex compartmentalized bureaucracies	● other means of data collection available ● treaty requires minimal uncoded signals ● no collection from 3rd country territory

2. SS-25 as a second new type ICBM	• potentially stabilizing with mobile basing and single warhead	• no impact with single warhead	no	• loose constructionist approach to complex treaty language	• SS-25 falls within salt compliant boundaries
3. SNDV ceiling exceeded	• no significant impact at SALT II levels	• no significant impact at SALT II levels	no	• no final agreements on bomber dismantlement/ conversion SOPs	• Soviet ceilings fall within agreed limits • US failure to ratify SALT II changes Soviet obligations
4. Krasnoyarsk radar	• significant only as part of a territorial defense	• contributes to more rapid deployment of territorial defense	yes	• definition influenced by logistical and geographic problems of construction	• implied guilt, Soviets willing to negotiate issue
5. launcher/ ICBM association	• no significance	• marginal at existing launcher/ warhead celings	yes	• vague issue–poor oversight of test SOPs	• agree on treaty provision/ may disagree test SOPs are a violation
6. activities at old SS-7 sites	• no significance at SALT II levels	• no		• considerable savings to use old infra structure for new mobile missiles	• disagree there is a violation. Define 'facilities' differently than US • issue is not applicable to mobile ICBMs

* These six issues were the confirmed violations listed in President Reagan's reports to Congress

Encryption of ICBM Test Data

Image One

Encryption is the coding of radio transmissions or telemetry from a test missile to its control stations.[18] These test data are a vital part of all research and development programs. They are essential to the testing party in determining whether a system meets operational requirements and to an arms control partner in verifying compliance with treaty commitments. Neither party wishes to reveal more information than is required by specific treaty provisions. In the case of SALT II, encryption is allowed except when it 'impedes' verification. How much encryption shields sensitive test data while communicating treaty compliant data is determined by the testing party. The failure of the SALT II Treaty to establish clearly how and when encryption impedes verification left an ambiguity that the Soviets have fully exploited.

The US has accused the Soviets of 'heavily' encrypting telemetry broadcasts during tests of their SS-24 and SS-25 ICBMs, as well as a new submarine-launched ballistic missile, the SS-N-20. These practices impede US verification of Soviet compliance with the SALT II Treaty. The President's reports to Congress concluded that Soviet concealment activities presented special obstacles to maintaining existing arms control agreements and undermined the political confidence necessary for concluding new treaties.[19]

Images Two and Three

The Soviets reportedly began encrypting signals from their missile tests in the mid-1970s, when they learned through espionage of the sophistication of US satellite monitoring capabilities. The technical capabilities for encryption devices demonstrated prior to SALT II required a long lead time for production and deployment aboard test missiles.[20] The issue from within the responsible Soviet bureaucracies was what modification would be required in Soviet encryption capabilities and practices after SALT II. Part of this difficult and technical question has to do with the relationship between legal encryption and the SALT II obligation not to 'impede' verification. Monitoring missile telemetry, like satellite photography, includes a high degree of overlap between its verification and intelligence gathering functions. What we would like to learn about Soviet missiles and what SALT II requires the Soviets to give us are two separate and often controversial issues.

The Soviets have never agreed that access to telemetry is necessary for verification. They maintain 'other' national technical means are capable of verifying the relationship between Soviet ICBMs and treaty obligations.[21] When this issue came up during the SALT II negotiations, the Soviets resisted. Eventually they gave in to US pressure, but with the proviso that it not require them to alter existing practices. American negotiators rejected this line noting that encryption was already hampering the intelligence monitoring (it became 'verification' after the Treaty) of provisions then under negotiation.[22] The final compromise was based on the limited data required to verify the new-type ICBM provision. Neither side would engage in 'deliberate' denial of telemetric information 'whenever' such denial impedes verification of compliance. A total ban on encryption was unacceptable to the Soviets. A specific list of data requirements would have opened far too much discussion of US intelligence techniques and procedures. The vague compromise in treaty language has resulted in persistent US charges of Soviet non-compliance.

The fact that the Soviets encrypt data to a substantial degree is troubling, but does not automatically constitute a violation. As the staff report to the SALT II Treaty pointed out, no criteria for determining when encryption impedes verification was ever established.[23] Nor have subsequent negotiations at the Standing Consultative Committee (SCC), established by the ABM Treaty as a forum for regular discussion and resolution of verification problems and disputes, been successful in working out specific procedures. If, for example, during a series of 20 ICBM test flights, data must be transmitted in the clear from all 20 tests? Some tests? How many tests? How much data? The SCC has become deadlocked on the issue. Soviet negotiators have asked US officials to explain what aspects of verification have been impeded. The United States has refused to present the Soviets with a model of treaty-compliant behavior, because such specifics might compromise US intelligence capabilities.[24] The dilemma is an exact replay of the original and unfortunate SALT II compromise that failed to specify what data was required to verify Soviet compliance.

The Soviet leadership is unlikely to modify existing test procedures as long as the US refuses to discuss the details of what it considers adequate compliance. It would be difficult for them to reconcile US demands with the position apparently taken by their technical specialists and the military that 'other means' of verification are available to determine the very selective performance data that Washington is entitled to have under existing treaty provisions.

An additional complication arises from the extensive deployment of US monitoring stations on the territories of its allies. The Soviet Union has never accepted third-country monitoring sites as legitimate 'national technical means'.[25] In practice this means that they need not respect limitations on interfering with the operations of such stations.

The SS-25 ICBM

Image One

In an attempt to limit the development and deployment of new, more capable ICBMs, the provisions of SALT II permit each side to test and deploy just one 'new type'. For the United States this was the MX; the Soviets designated the SS-24 as their new missile. By almost any definition except one the SS-25 ICBM is a second new missile, and not, as the Soviets claim, a permissible follow-on to the SS-13. The *one* exception may be the SALT II definition of 'new type'.

This compliance controversy has grown out of a combination of complex treaty provisions and inadequate monitoring capabilities for those provisions. For example, both sides wished to maintain the right to modernize their existing missile inventories–new warheads, improved guidance systems. Exactly when a new or modernized missile became a 'new type' was spelled out in detail. Treaty language defines a 'new type' as any missile that would change an existing missile's fuel type; number of stages; or, by 5 per cent or more, its length, largest diameter, launch weight, or throw weight. An additional provision precludes the deployment of a new single warhead ICBM if the warhead weight is less than half of the overall throw-weight. The intention of the latter provision was to prevent the sudden emergence of multi-warhead missiles.

When the US first detected testing of the missile in early 1983, American intelligence concluded that it was a 'new type' based on estimates of its dimensions, throwweight, and warhead weight. US officials concede, however, that their data base on the SS-13 is poor and incomplete.[26]

Images Two and Three

The first problem that Image One or Two analysis encounters with the SS-25 ICBM issue is establishing an accurate accounting methodology. Administration reports claim:

- The SS-25 is an illegal second new type ICBM;
- The weight of its warhead is less than 50 per cent of its total throwweight (thereby posing an illegal threat of a MIRV capable ICBM);
- The Soviets have encrypted vital data during its test flights;
- The Soviets conceal the relationship between the SS-25 ICBM and its launcher;
- The Soviets have used 'remaining facilities' at former SS-7 ICBM sites to support SS-25 activities.

Does the SS-25 represent one or five violations? Or is it one issue that violates five provisions of SALT I and II? The most serious issue is the missile's legality as a modernization or follow-on to the SS-13 rather than a second, illegal new type ICBM. This issue rests on a controversial Soviet definition of throwweight. They have argued at the Standing Consultative Committee in Geneva that a device, which normally would be considered part of a missile's throwweight, is attached to the SS-13 missile's third stage. The United States has not included this device in its estimates of SS-13 throwweight. The results have been consistently undervalued throwweight for the SS-13 which, in turn, has produced an SS-25 data base that exceeds the SALT II 5 per cent provision for ICBM modernization. The Soviets, as a result of this alleged SS-13 testing procedure, claim that the throwweight of the SS-25 is actually less than that of the SS-13.

The second SS-25 violation, the testing and deployment of a single re-entry vehicle that weighs less than 50 per cent of an ICBM's total throwweight, has also been denied by the Soviets. They have argued that during testing, they have attached a heavy instrumentation package that increases the missile's throwweight, and thereby decreases the re-entry vehicle's percentage of the total throwweight. According to the Soviets, when a deployed SS-25 is tested without this package the missile complies with the treaty.[27] The Soviets commonly test operational missiles and, depending on their degree of encryption or alternate US monitoring capabilities, this part of the controversy could eventually be resolved.

Krasnoyarsk Radar

Image One
Krasnoyarsk could be remembered by historians as the central Siberian birth place of the late Soviet leader, Konstantin Chernenko.

Instead, its notoriety in the West is the result of the Krasnoyarsk Radar Station–'Exhibit A' and consensus violation of the ABM Treaty.

Discovered by US intelligence in 1983, the radar's single face looks across 6000 miles of Soviet territory toward the northern Pacific. When completed, the radar will have the appearance and capability of other Soviet early warning radars already operational along the Soviet periphery. These radars comply with the ABM Treaty because they are located on the edge of Soviet national territory and are sited outward. The Krasnoyarsk location, if not its function, however, violates the intention of the ABM Treaty to restrict radars–especially large phased-array radars capable of both legitimate early warning and prohibited ABM battle management missions. Restricting radar locations was the best solution to the problems of limiting a technology with both legitimate and proscribed missions. Radars are especially sensitive ABM components because they play a critical role and require a long lead time for construction if the Soviets planned to develop a territorial defense in violation of the ABM Treaty.

Image Two and Three
The Soviets have countered US claims of non-compliance by asserting that the radar is a treaty compliant space-tracking station and this will become clear to US observers when the radar begins operations. The issue underlines the great difficulties in negotiating functional limitations on versatile technologies. Even so, the Soviets have tacitly admitted doubts on the strength of their position by offering to stop construction at Krasnoyarsk if the United States will do the same at Thule, Greenland, the site of a nearly 30-year-old early warning radar currently undergoing extensive modernization which some former US officials have argued is also a violation of the ABM Treaty.[28] Whether the Soviets have raised the US radar at Thule as an analogy to defend themselves or to find a face-saving compromise is uncertain. The larger issue is why and how the decision was made to locate the radar at Krasnoyarsk.

Krasnoyarsk fills the last gap in the Soviet early warning radar system. Difficult lessons may have been learned from Soviet experience with earlier radar construction. Building along the northern periphery of the Soviet Union may have been more expensive and more difficult than imagined. They may also have discovered that in actual operation, these radars had a narrow field of vision. Filling the

Can We Make This Trip With the Russians? 117

last gap in northeastern Siberia may have required two radars distant from existing rail or road networks. A decision to build on the periphery of the existing Soviet logistical support base could have seemed appropriate, even logical to those involved during the declining years of Brezhnev and the ascendancy of the Reagan administration.

Even the most extreme loose constructionist model of Soviet compliance decision-making has difficulty in accommodating an expenditure as large as Krasnoyarsk. Nevertheless, if the conditions described above are valid, the gambit may have seemed worth the risk, particularly if it were not accompanied by a larger scheme to break out of the ABM Treaty. Any such long-range plan, if it existed at all, would most probably have been devised within the bureaucratic subculture (Party-military hardliners, for example) rather than by an enfeebled Politburo undergoing the first of three leadership successions in almost as many years. The ABM Treaty, however, makes no provisions for such practical problems or internal political friction. The radar is the most convincing issue in the President's non-compliance reports.

Exceeding The Strategic Nuclear Delivery Vehicle (SNDV) Limits of SALT II

Image One
This complex issue arises from the logical and reasonable assumption that the 1982 mutual pledge to abide by the major provisions of the SALT II Treaty includes some numerical ceiling on the deployed numbers of strategic nuclear delivery vehicles. US ratification of the Treaty, however, was the only means of activating the specified SNDV ceilings. Without ratification, no explicit agreement exists on the numerical value of a launcher ceiling.[29] The US has used 2504 as the aggregate number to which the Soviets have politically committed themselves, because that number was in effect in 1979 when the SALT II Treaty was signed. In its December 1985 non-compliance report, the administration concluded that the Soviets were in violation by exceeding that number. The administration's case is based on Soviet activities between September 1981, when Washington informed the Soviets that it would not seek to ratify SALT II, and the 1982 mutual political commitment to abide by the Treaty. During this period, the Soviet Union deployed additional SNDVs, and those activities continued through 1984 and 1985, including the destruction

of older SS-11 launchers and silos as new SS-25s were deployed. *Bear H* cruise missile-carrying bombers were also deployed, but older bombers were not destroyed in sufficient numbers to compensate. The net effect was to maintain their SNDV count at a level slightly above the 2504 ceiling claimed as binding by the United States.[30]

Images Two and Three
Whether the Soviet Union has violated the implicit SALT II limits on strategic nuclear delivery vehicles depends on how the data is interpreted for strategic bomber forces. The procedures for dismantling or converting bomber aircraft to aerial refueling tankers have been drafted, but not completed by the Standing Consultative Commission. The lack of specific agreement on bomber dismantling procedures, combined with uncontested Soviet compliance with the dismantlement of other strategic weapons, raises serious questions about the strength of the US case. The President's reports fail to note that as new Soviet strategic systems have been deployed, older systems (nearly 1300) have been retired.[31] Missile silos have been destroyed; missiles, and submarines have been cut up according to SALT II procedures.

The Joint Chiefs of Staff have also reported that the tail sections of 15 *Bison* bombers were cut off as the Soviets deployed new *Bear H* cruise missile carriers.[32] At issue are approximately 30 Bison bombers, which were converted to aerial fuel tankers, but which the US still counts as operational bombers.[33] The Soviets, while not publicly stating their case, may consider these newly converted tankers to be no longer SALT-accountable and in the same category as other Bison tankers that were formally exempted from SALT II limits. Even with the unresolved tanker issue, the administration claims that the Soviet SNDV ceiling has varied between 2504–2540 during 1985, with the most recent figure set at 2520–16 over the US claimed ceiling.[34] If the drafted bomber-tanker conversion procedures were signed at the SCC, the SALT accountable Soviet limits would drop to 2490 or 14 below their permitted limits. If an alleged violation can be resolved at the SCC through formal agreement on bomber dismantlement and conversion procedures already in draft form, its credibility in the Image One cheating scenario is substantially reduced.[35]

Launcher and ICBM Association

Image One
SALT II prohibits 'deliberate concealment measures which impede verification by national technical means', including activities that conceal the association between an ICBM and its launcher during testing. Is, for example, a missile being launched from a silo or a mobile launcher? Precision is required for counting launchers, identifying them, and differentiating them from launchers of missiles that are not limited by the Treaty–SS-20 mobile, intermediate-range missiles for example.

The administration's December 1985 compliance report added this issue for the first time. Without spelling out the details, the report concludes that the SS-25 test series have violated SALT II provisions by concealing the relationship between the missile and its launcher.[36] A statement by the Soviet Government that the SS-25 has a mobile launcher has not been judged sufficient to relieve the Soviets of their obligation to allow the United States independent access to make its own judgment on the issue.

Image Two and Three
This is the fourth issue involving the SS-25 ICBM. The concealment of the relationship between the missile and its launcher would have potential military significance if the Soviets had declared their intent to deploy the SS-25 in silos. A case could then be made that perhaps the Soviets might attempt covert deployment of mobile missiles. However, since the Soviets have not only declared, but have already deployed the SS-25 on mobile launchers, which are presumably distinguishable from other Soviet mobile launchers, the issue has no military relevance.

The degree to which Soviet organizational routines at the SS-25 test range may violate concealment prohibitions in SALT II represents a principle worth calling to the attention of Soviet leaders; but it is not an issue that would have likely been directed or controlled from the top. Image two offers the most likely explanation of Soviet actions. Responsible bureaucrats may have attempted to conceal something. They could just have easily assumed that US intelligence could discriminate between mobile launchers and silos.

120   Deterrence and Defense in a Post-Nuclear World

The Use of 'Remaining Facilities' at Former SS-7 Sites

Image One
Soviet activities at some former SS-7 ICBM sites became the single SALT I issue to be upgraded from 'no violation' to 'violation' in the December 1985 report. SALT I prohibited the use of certain facilities remaining at dismantled or destroyed ICBM sites. This prohibition was devised to prevent the rapid reactivation of old sites. Construction activity during 1984 and 1985 at retired SS-7 sites was judged to be incorporating remaining facilities in support of the deployment and operation of new SS-25 mobile ICBMs.

SALT I and the procedures for dismantling ICBMs negotiated in the SCC established two major criteria for determining whether the use of a structure remaining at any former SS-7 sites would be a violation. For this purpose, 'structure' must meet the definition of a 'facility', and the 'facility' must be used for storage, support, or launch of ICBMs.[37]

The administration's December 1985 compliance report and the 1986 ACDA report found the Soviets in compliance with their destruction and deactivation of SS-7 missiles and launch sites. However, both reports concluded that a 'number' of surviving facilities were being used in support of SS-25 mobile ICBMs.

Image Two and Three
This sixth and final issue related to the SS-25 ICBM is the one issue most clearly explained by Image Three. There is a sharp disagreement between the parties over this SALT I issue. The Soviets maintain that any remaining facilities at former SS-7 ICBM sites used in support of SS-25 activities are legal because the provision did not apply to mobile missiles. Even when mobile missiles are brought into the debate, it can be argued that the SS-25 is being deployed in self-sufficient support bases similar to SS-20 intermediate range missiles. Remaining SS-7 facilities are treaty compliant and support construction and logistical functions for the SS-25 bases, but not for 'storage', 'support', or launch of ICBMs. Even if the issue were more ambiguous than it seems to be, powerful forces in the Soviet bureaucracy would find it attractive to cut costs by integrating as much existing infrastructure as possible into Soviet strategic modernizations programs.

THE MILITARY SIGNIFICANCE OF NON-COMPLIANCE ISSUES

The available evidence supports conclusions that vary significantly from official non-compliance reports to Congress. One important test of Images Two and Three developed above is to assess the military significance of each charge. Is there a pattern? Does that pattern suggest a well orchestrated strategy of non-compliance to gain unilateral military advantage over the United States? If Soviet activities have resulted in no significant military advantages, the case for alternative interpretations of the evidence becomes more compelling.

Table 4.3 summarizes the military significance of the six 'confirmed' violations. Together, these violations suggest no clear pattern that strengthens Soviet strategic posture in any one of the three criteria for significant military behavior. Soviet behavior can be explained through a combination of political, technical, and legal motivations. The violations described here are such blatant acts that they cannot be called 'cheating' in the usual sense of clandestine activities to gain a military advantage or strategic surprise. Even if the Soviets were exploiting the plausible denial that often results from vague treaty language and imperfect monitoring capabilities, the military benefits are marginal in comparison with the political damage to overall Soviet-American relations.

For example, the long-term military significance of encrypting missile flight test data depends largely on the degree to which 'other' intelligence methods – including ground, sea, or space based radars – can, as the Soviets imply, gather required data. Telemetry is the most reliable single means of making firm estimates, and total encryption would undermine US ability to predict the direction of Soviet strategic modernization programs. On the other hand, it is difficult to reconcile US claims that encryption has impeded verification of treaty selected performance characteristics with the detailed official descriptions of Soviet missiles, and with the precise data required to refute Soviet claims that the SS-25 is a legal follow-on to the SS-13 rather than a 'new type' ICBM. Secretary Weinberger, for example, has described the SS-25 in precise terms–it is 10 per cent longer, 11 per cent larger in diameter, and has 92 per cent more throwweight than the SS-13.[38]

Technical and operational procedures seem to be at the heart of the SS-25 issue. Recent Soviet activities are difficult to explain in

TABLE 4.3 *Military Significance of Soviet Non-Compliance*

Issues*	Strategic balance/stability	Breakout potential	Undermines predictability
1. Encryption of test data	– no significance at SALT II levels	– marginal depending on other monitoring capabilities	yes
2. SS-25 as a second new type ICBM	– potentially stabilizing with mobile basing and single warhead	– no impact with single warhead	no
3. SNDV ceiling exceeded	– no significant impact at SALT II levels	– no significant impact at SALT II levels	no
4. Krasnoyarsk radar	– significant only as part of a territorial defense	– contributes to more rapid deployment of territorial defense	yes
5. Launcher/ICBM association	– no significance	– marginal at existing launcher/warhead ceilings	yes
6. Activities at old SS-7 sites	– no significance at SALT II levels	no	no

* These six issues were the confirmed violations listed in President's Reagan's reports to Congress

terms of duplicity or deception. Why, for example, in the midst of a controversy over the SS-13/25 relationship would the Soviets resume test flights of old SS-13s if they were deliberately cheating or attempting to obscure the technical relationships between the two missiles?

From the former administrations's point of view the Soviets have violated the SALT II limitation of one new missile type. Ironically, the military significance of the SS-25 may be in the interest of both the United States and the Soviet Union. Strategic modernization trends toward less vulnerable, smaller, single-warhead ICBMs have been advocated by many US Government officials and defense analysts, including the Scowcroft Commission.[39] Washington should be pushing the Soviets in precisely this direction, just as many are pushing for a new US mobile missile–the Midget Man. Strategic stability in the form of fewer incentives to launch a first-strike could be strengthened by mobile ICBM deployments in combination with reductions in the Soviet SS-18 ICBM force.

Perhaps the most significant SS-25 issue is the throwweight of its warhead. If, as the US claims, it is less than 50 per cent of the total weight of the re-entry vehicle, a limited 'breakout' capability would result in the capacity to deploy a multiple-warhead version. Such engineering and packaging games would not affect the strategic balance unless the Soviets attempted to deploy forces above SALT II limits or until arms control agreements were to bring strategic nuclear force levels down to at least 50 per cent below their current levels. Even then, the survivability of US forces would depend as much on their deployment modes as on marginal numerical advantages by either side. In any event, there will be adequate time for the US to respond if future Soviet tests of deployed SS-25s confirm the American throwweight estimates.

Two additional SS-25 issues described in the President's reports seem less serious. The military significance of Soviet efforts to conceal the relationship between the SS-25 missile and its launcher would have potential military significance if the Soviets had declared their intent to deploy the SS-25 in silos. A case could then be made that perhaps the Soviets might attempt covert deployment of mobile missiles. However, the Soviets have not only declared, but have already deployed the SS-25 on mobile launchers which are distinguishable from other Soviet mobile launchers. The issue has little military relevance, but left unresolved, a precedent could be established that undermines the US capability to predict the direction of Soviet strategic modernization programs if future Soviet test practices do not clearly show US intelligence a single launcher and missile association.

The military significance of Soviet actions at former SS-7 ICBM sites is negligible as long as they remain within SALT I and SALT II ceilings. The only measurable advantage is the savings which result from using existing roads and facilities to support SS-25 bases.

There is far less ambiguity in Soviet radar deployments. Krasnoyarsk is an illegal radar site. The best test of its military potential is to assess its capabilities against the deliberate Soviet ABM breakout thesis. There are at least four functions that Krasnoyarsk could perform: early warning (legal), satellite tracking (legal), ABM battle management (illegal), and ASAT (anti-satellite) battle management (illegal). According to a 1985 CIA assessment, the radar is not well designed for ABM battle management because its single face (unlike the legal four-sided battle management radar near the Moscow ABM complex) looks in the wrong direction to detect the northern trajectories of US ICBMs and their warheads against which the radar would have to guide interceptor missiles.[40] Moreover, the radar is not 'hardened' against nuclear effects, and there are no interceptor missiles, associated radars, or other ABM-related items near the facility.

The CIA report, as well as former Secretary of Defense Harold Brown, conclude that the radar is a violation, but not a military threat.[41] In fact, the administration's non-compliance reports indirectly make the same case through their inability to confirm that other suspected ABM violations–such as mobile components and dual capable air/missile defense radars–are valid, much less linked in any way to the Krasnoyarsk facility. Cheating in the gray areas between air defense and ballistic missile defense is risky, because it results in low confidence defenses while risking offensive countermeasures. The case for a creeping Soviet territorial defense or breakout from the ABM Treaty requires clear evidence of a more systematic effort, including hundreds if not thousands of radars, sensors, and ICBM interceptors, all of which require long lead times for construction and deployment as part of a credible territorial defense.

Finally, and by all accounts, the issue of Soviet strategic nuclear delivery vehicles in excess of SALT II ceilings has little military significance, and does not, when combined with the destruction patterns of Soviet strategic forces and the unresolved bomber/tanker conversion procedures, support the case for concerted cheating to gain strategic advantage. The numbers in dispute rest on the bomber conversion issue. These bombers, when converted to refueling tankers, will support a modernized Soviet bomber force, but their numbers are too small to undermine the predictability of Soviet force modernization or to challenge the strategic balance.

CONCLUSIONS

The marginal military gains from Soviet activities described under Image One and other plausible models of Soviet behavior assessed under Images Two and Three support the conclusion that the Reagan Administration's attacks against the major pillars of the existing arms control regime–SALT I, SALT II, and the ABM Treaty–exceed the principle of proportionate response. At the same time, those in the United States who have consistently argued the case for Soviet non-compliance are aided by Soviet counterparts who are equally anxious to exploit every treaty ambiguity. The result is a political climate in which images of Soviet cheating and actual Soviet bureaucratic behavior are natural allies in fomenting protracted compliance debates that threaten the Soviet-American arms control regime.

If the image of Soviet bureaucratic politics and standard operating procedures described here has even marginal validity, the Standing Consultative Committee has been an ingenious device to cut through the highly secretive and compartmentalized Soviet bureaucracy, confronting political leadership formally and directly with US concerns about compliance behavior. The SCC functioned in this manner during the Nixon, Ford, and Carter administrations.[42] This is not to suggest that all issues were resolved smoothly or that the Soviets, when confronted with US concerns, pleaded guilty or confessed. Behavior, however, did often change over time, demonstrating a potential relationship between the SCC and the Soviet leadership's active intervention to alter bureaucratic routines that were at odds with US interpretations of SALT II and the ABM Treaty.

The Reagan Administration was far more critical of the SCC. The number of non-compliance issues has risen dramatically along with a much greater tendency to discuss them publicly rather than at the SCC. SCC negotiators have experienced a greater degree of frustration and more dead ends, as in the case of encryption discussed above. Its function as a continuing negotiating forum for resolving SALT and ABM Treaty disputes will produce fewer visible results when and if Image Three (differences in interpretations of treaty obligations) or Image One (pre-planned deception to gain military advantages) are at work. Image Three requires complex and often protracted negotiating that matches that of the original treaty. Image One, to which the Reagan Administration gravitated, threatened not only the SCC, but the entire arms control regime–present and future.

Soviet behavior has not been the only critical variable. Powerful forces in the administration repeatedly made it clear during Reagan's first term that strategic modernization programs ('supply-side arms control' as former Assistant Secretary of Defense, Richard Perle, once quipped) were the top priority. The Strategic Defense Initiative introduced another program that US negotiators were required to protect at the negotiating table. Support for both offensive and defensive modernization programs was made easier by the administration's frontal assault against Soviet non-compliance with their arms control obligations.

There are important lessons in these events if arms control is to have a role in future Soviet-American relations. For the Soviets it would be a mistake to dismiss all of the allegations against their compliance record as propaganda from unregenerate hawks. The Reagan Administration struck a sympathetic nerve in the American public, because mistrust of the Soviet Union is widespread, if not a fundamental part of American beliefs. Arms control, even though widely supported, always walks a fine line between public fears of nuclear war and Soviet intentions. Loose constructionist compliance behavior will have to give way to something closer to the US legalistic or tight constructionist views of treaty obligations if arms control is not to be supplanted by unilateral military responses to Soviet threats.

The Soviets can respond to US concerns only if Washington establishes politically realistic standards of compliance. Unfortunately, American domestic politics has produced three competing standards.[43] First, within the most enthusiastic wing of the arms control community there is often an implicit acceptance of ambiguity or even of 'minor' violations as long as they have no military significance. By contrast, legalists argue that all violations are important, and the degree to which parties adhere to the provisions of a treaty is an important measure of their good will and trustworthiness. Precise language strictly adhered to is fundamental to treaties between States.

The Carter Administration was split by these two factions just as the Reagan Administration was divided between legalists and a third and increasingly dominant faction–the skeptics for whom no degree of arms control verification can buy more security than unilateral defense programs. For skeptics, suspicion is the basis for action. The Soviets must prove the absence of non-compliance, because even compliance with provisions of a treaty can be interpreted as sinister. The skeptics' influence can be found in many of the ambiguous conclusions of the administration's compliance reports. The most recent example is the internal debate that followed the discovery of three new, but treaty compliant radars discussed above. Though these radars were located at legal sites, several administration officials wanted their existence cited as evidence of a Soviet program to construct an illegal territorial ABM defense system.

The loose constructionist school is politically unacceptable in the United States. On the other hand, the skeptics demand standards of evidence that could be satisfied only by levels of intrusive verification that are as unacceptable to US negotiators as they are to the Soviets. By default, the legalist standard survives as a measure most likely to be accepted by both American and Soviet officals. It is also the standard best suited to deal effectively with all three images of Soviet compliance behavior.

Precise treaty language, reduced levels of complexity, and ambiguities resolved by more intrusive means of verification are reasonable standards to incorporate in future treaties. The Soviets surprised the Reagan Administration by accepting several complex provisions for on-site inspection. In recent negotiations on the elimination of chemical weapons the Soviets accepted surprise or challenge inspections of sites where a potential treaty violation is suspected. The INF Treaty reveals how extensively the Soviets agreed to US proposals for on-site inspection of production facilities, deployment areas, and during the destruction of missiles and support facilities covered by the treaty.[44]

Provisions for on-site inspection raise confidence levels and remove obstacles to ratification in Congress, but they are not foolproof. Neither are they likely to satisfy skeptics. With the possible exception of the SS-16 ICBM, on-site inspection would not have resolved any of the compliance issues discussed here. Cheating at the margin will always be possible, but cheating to the degree that one side gains a significant military advantage before the other responds is difficult and risky. Soviet nuclear weapons are not produced by cottage industries. Their components, production, testing, and deployment require massive infrastructures that are visible to US intelligence (see Figure 4.1). For the Soviets to break out of treaty obligations with programs that suddenly and dramatically alter the strategic balance would require something close to a total intelligence failure.

The prospects for intelligence failures, no matter how slight, have been and should always be hedged by robust research and development programs which provide both insurance against cheating and incentives not to. Russian history from Potemkin village to the

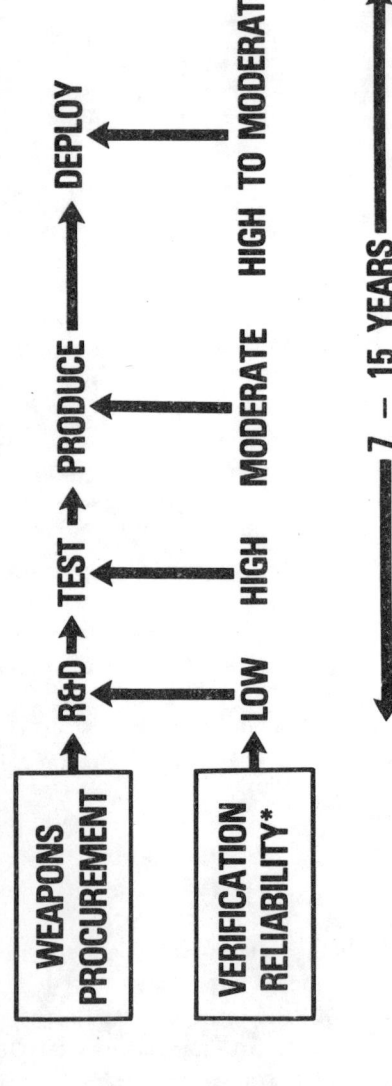

FIGURE 4.1 *Weapons' Cycle and Verification*

bomber and missile gap controversies of the late 1950s and early 1960s is filled with examples of deception to conceal weakness, not strength. Military strength is the one symbol on which the Soviets have rested their superpower status. They wear it proudly for all its cost and sacrifice. Examples of strategic surprise that have revealed Soviet strength are limited mostly to wartime experiences, such as the battle of Stalingrad, which the Soviets won by counter-attacks by massive reserves unknown to the Germans. In war, however, deception need only work once for a short period. Protracted deception required for breaking out of an existing arms control regime with significant military capabilities is much more difficult and extremely dangerous when compared to the political and military benefits the Soviets could gain through a more stable strategic environment that results from arms control. Cheating also wastes the considerable political capital the Soviets have invested in the process of negotiating treaties in the first place.

Multiple explanations for Soviet actions and marginal military benefits from the Reagan Administration's cheating scenarios did not justify the decision to abandon SALT and threaten the ABM Treaty. The danger in these trends stems from long gaps between old and new treaties, which may be filled by destablilizing offensive of defensive actions by one or both parties. An arms control failure at the end of costly strategic modernization programs and in spite of a strong negotiating posture would have left the Reagan Administration with a dubious legacy and confronted President Bush with formidable obstacles on the path of Soviet-American relations. Obstacles remain to a new strategic arms treaty, but the INF Treaty and Soviet concessions on on-site inspection have made them considerably less intractable.

NOTES

1. Described in Raymond L. Garthoff's, *Detente and Confrontation: American-Soviet Relations from Nixon to Reagan* (Washington, DC: Brookings, 1985), p. 300.
2. V. Issraelyan, *Final Record of the 119th Meeting,* Committee on Disarmament, Document CD/PV. 119, Geneva, 31 March 1981, pp. 16–17. For detailed analysis of Soviet verification perspectives see Stuart A. Cohen, 'The Evolution of Soviet Views on SALT Verification: Implications for the Future', in William Potter (ed.), *Verification and SALT* (Boulder, Co: Westview Press, 1980), pp. 49–76, and Allen

Krass, 'The Soviet View of Verification', in William Potter (ed.), *Verification and Arms Control* (Lexington, Ma.: Lexington Books 1985), pp. 37–62.
3. Quoted in Thomas Wolfe, *The SALT Experience* (Cambridge, Mass: Ballinger, 1979), p. 261.
4. Evidence of the SS-25 became available by 1983; the Krasnoyarsk Radar in 1983; and encryption practices were common prior to SALT II, stopped, but resumed again with tests of the SS-25 and the SS-NX-20–a sea-based missile. The Soviet radar was well under way when first discovered. Construction probably began in the 1980–81 time frame. This is not considered to be an intelligence failure. With finite resources, it is axiomatic that one will not find what one is not looking for.
5. Soviet sources claim that the Krasnoyarsk radar is part of a nine-installation project planned in the late 1960s and approved in the early 1970s. Soviet defense officials constructed the first eight radar complexes at sites near their borders as required by the ABM Treaty. The Krasnoyarsk site was selected to save money. Former Defense Minister Dimitri Ustinov, according to these sources, did not tell the Politburo that the radar potentially violated the treaty. Actual construction at the site did not begin until early 1982, and was discovered by US intelligence in July 1983. See *Washington Post*, 14 December, 1987, P. A17.
6. Scilla McLean (ed.), *How Nuclear Weapons Decisions Are Made* (New York: St. Martin's, 1986), Chapter 1.
7. According to Roy Medvedev, the Defense Council was created in 1958 to unify the leadership of all 'war departments'. Roy Medvedev, *Khrushchev* (Garden City, New York: Doubleday, 1983), p. 128.
8. Nikita Khrushchev, *Khrushchev Remembers: The Last Testament*, editor and translator, Strobe Talbot (Boston: Little Brown, 1971), pp. 41–2, 50.
9. Condoleezza Rice, 'The Party' The Military, and Decision Authority in the Soviet Union', *World Politics* Vol. XL, No. 1, October 1987, p. 75.
10. The important role of the military at SALT I is illustrated by American accounts of the rigorous reaction shown by Soviet military members of the delegations when US negotiators gave their civilian counterparts data on Soviet strategic forces. That information was not to be shared with unauthorized civilians.
11. Speculation about succession began in 1975 following Brezhnev's stroke and disappearance from public life for several months. Clear signs of ill health were present in 1977, a period when he had 'packed' the Politburo with close friends, increasing its size from 11 to 14. As Brezhnev's health deteriorated further in 1978, speculation in the foreign press about his successor began in earnest, continuing as something of a cottage industry in the West until his death in November 1982. See Zhores A. Medvedev's, *Andropov* (New York: W.W. Norton, 1983), pp. 3–6.

12. Described by Stephen M. Meyer, 'Civilian and Military Influence in Managing the Arms Race in the USSR', in Robert J. Art, Vincent Davis, and Samuel P. Huntington's, *Reorganizing America's Defense* (Washington, DC: Pergamon-Brassey, 1985), p. 55.
13. Ibid. p. 53
14. Charles H. Fairbanks, Jr., 'Bureaucratic Politics in the Soviet Union and in The Ottoman Empire', *Comparative Strategy*, Vol. 6, No. 3, 1987, pp. 351–2.
15. See, for example, James. R. Millar, *The ABCs of Soviet Socialism* (Chicago, Il.: University of Chicago Press, 1981), Chapters 6–8.
16. The President's Reports to Congress cover compliance issues for all Soviet-American arms control treaties in force. Only SALT I, SALT II, and ABM Treaty issues are examined here. A fifth non-compliance report sent to Congress in December 1987 is not included in this study. Its findings are essentially the same as the March 1987 report.
17. Described by Michael Gordon, *New York Times*, 19 December 1986, p. 6.
18. For a detailed study of the technical aspects of monitoring arms control compliance see Allan S. Krass, *Verification: How Much Is Enough?* (Lexington, Mass.: Lexington Books, 1985).
19. *Soviet Non-Compliance*. (Washington, DC: US Arms Control and Disarmament Agency, February 1986), p. 11.
20. Krass, op. cit., p. 188 and note 28.
21. The author is grateful to Arnold Horelick on this point. See also, *Explaining Soviet Compliance Behavior: Conference Digest*, Stanford University, 14 February 1986 (Mimeo), p. 53. For discussion of 'other ways' to derive data on Soviet ICBMs see Krass, op. cit., p. 189.
22. James A. Schear, 'Arms Control Treaty Compliance: Buildup to Breakdown', *International Security*, Vol. 10, No. 2, Fall 1985, p. 163.
23. John Pike and Jonathan Rich, *Soviet Compliance with SALT II and the ABM Treaty*, Report by the Federation of American Scientists, 7 March 1984, p. 7.
24. Described by Jack Mendelsohn, 'Reagan's Proportionate Response Policy: Sense or Nonsense?', *Arms Control Today*, Vol. 16, No. 1, January/February 1986, pp. 7–11.
25. Krass, op. cit., p. 187 and note 28.
26. Schear, op. cit., p. 166, and Pike and Rich, op. cit., pp. 8–10.
27. Mendelsohn, op. cit., p. 8.
28. Legal and technical issues are described by former ACDA staff member Peter Zimmerman in 'The Thule, Fylingdales, and Krasnoyarsk Radars', *Arms Control Today*, Vol. 17, No. 2, March 1987, pp. 9–11.
29. The Soviet Union declared that as of the date of signing SALT II it possessed 1398 ICBM and 950 SLBM launchers and 156 heavy bombers. Since the SALT II levels of 2400 and 2250 are applicable only 'upon entry into force' of the treaty, this total of 2504 is considered to be the implicit limit on Soviet SNDVs.
30. *Soviet Non-compliance*, op. cit., pp. 8–9.

31. *Joint Chiefs of Staff Military Posture Statement, FY 1987* (Washington, DC: US Department of Defense, 1986), p. 19. Chart is compared with FY86 issue.
32. Ibid., p. 19.
33. *New York Times*, 24 November 1985, p. 18.
34. *Soviet Non-compliance*, op. cit., p. 9.
35. The administration confirmed in a letter to Congressman Lee Hamilton that no effort has been made to complete bomber dismantlement/conversion procedures at the SCC. This raises an important consideration: how much was the Reagan Administration responsible for creating the Soviet record of compliance?
36. *Soviet Non-compliance*, op. cit., pp. 11–12. One can only speculate on the specifics of this issue. A reasonable scenario is the deployment of an SS-25 mobile launcher near a test silo or launch pad, resulting in US monitoring uncertainties about which launching mode was actually used.
37. Ibid., p. 13.
38. Secretary of Defense Weinberger, during a briefing for his fellow NATO defense ministers in Brussels. *Los Angeles Times*, 30 October 1985, p. 16.
39. Scowcroft, B. *et al.*, *Report of the President's Commission on Strategic Forces* (Washington, DC: US Government Printing Office, 1983), pp. 9–21.
40. See Michael Gordon's Report, 'CIA is skeptical that New Soviet Radar Is Part of an ABM Defense Report', *National Journal*, Vol. 17, No. 10, 9 February 1985, pp. 523–6.
41. Ibid., p. 525.
42. Questions about Soviet compliance began surfacing publicly in 1975. According to a 1978 Department of State Report to Congress, the US raised eight different issues in the SCC, six of which were resolved. Two other issues were unresolved due to ambiguities in the agreements themselves. See Mark M. Lowenthal and Joel S. Wit, 'Politics, Verification and Arms Control', *Washington Quarterly*, Vol. 7, No. 3, Summer 1984, 1984, pp. 117, 125.
43. The author is indebted to Glenn C. Buchan for these typologies. See his 'The Verification Spectrum', *Bulletin of the Atomic Scientists*, Vol. 39, No. 9, November 1982, pp. 16–19.
44. The INF verification regime is spelled out in the treaty text, articles IX–XIII and in the 'Inspection Protocol' to the Treaty. See also Jack Mendelsohn, 'INF Verification: A Guide for the Perplexed', *Arms Control Today*, Vol. 17, No. 7, September 1987.

5 Security In a Post-Nuclear World: The Unfinished Agenda

> I consider the protection of peace and freedom with these annihilating weapons to be very problematic. The bombs fulfill their purpose only if they never fall. But if everyone knows that they will never fall, they do not fulfill their purpose.
> German Physicist Carl von Weizaecker.

The preceding chapters identify trends toward a post-nuclear stage in the superpower strategic relationship. Nuclear weapons are not yet obsolete, but they may become less important to our security. This is in part the result of the failure to develop nuclear deterrence and nuclear war fighting into a single coherent strategy. Equally important are new technologies that are emerging from the Strategic Defense Initiative and from research and development programs for advanced conventional weapons. Together, they have the potential to make nuclear weapons comparatively less important instruments of deterrence, and to push military strategy further away from its nuclear dependency.

These trends can be seen in the strategic innovations of the Reagan years. During the 1970s, American strategy for nuclear deterrence became increasingly offensive. The ABM Treaty insured mutual vulnerabilities to nuclear attack, codifying the axiom that the best defense is a good offense. During this same period, NATO strategy for conventional war in Europe became increasingly defensive in character. Policy and doctrine repeatedly stressed NATO's defensive mission and rejected offensive military operations.[1]

President Reagan set forces in motion intending to reverse the offensive character of nuclear deterrence and the defensive cast of conventional strategy. At the strategic nuclear level, arms control and strategic defense attempted to reverse a long-standing strategic orthodoxy that emphasized the importance of offensive nuclear weapons to deter war or deny victory to the enemy. In Europe, strategy and force modernization during the Reagan years took on a decidedly offensive posture. 'Deep strikes' against Warsaw Pact territory, the Army's Airland Battle doctrine with its emphasis on seizing the initiative and engaging in offensive actions, and the Navy's maritime strategy for aggressive forward deployment are examples of offensive strategies that rely on conventional weapons for their credibility.

The Reagan Administration set the stage for a post-nuclear ea. But its legacy is flawed by a lack of coherence among three of its most important strategic programs–arms control, modernization of offensive nuclear weapons, and strategic defense. For these programs to merge in the co-ordinated pursuit of stable deterrence, the new administration must first reconcile the contradictory goals that have been declared for each.

ARMS CONTROL STRATEGY AND NUCLEAR MODERNIZATION: SHAPING DETERRENCE STABILITY

Arms control treaties provide a framework for achieving strategic stability–that ideal posture where neither side has an incentive to initiate a nuclear attack. By contrast, strategic instability is a posture where neither the United States nor the Soviet Union believe that 'victory' can be achieved or defeat averted without striking the other side pre-emptively. This state of disequilibrium provides mutual crisis stability only if offensive forces are reduced to levels sufficiently low that retaliatory forces can be maintained through unilateral actions. Early warning systems, high alert rates, and mobility of weapons are the most obvious examples that reduce incentives for attacking first.

Stability can create an environment in which both sides feel less compelled to deploy new offensive or defensive weapons that offset the deployments of the other. Proponents of strategic defense and offensive pre-emption reject stability based on offensive forces that deter by threat of retaliation. The traditional model of deterrence is, they argue, based on revenge, not defense.

The introduction of strategic defense complicates the offensive-defensive relationship by adding a new element of competition. This competition is based on the prudent expectation that no weapon is purely defensive. One-sided defensive advantages can support offensive threats. Offensive threats backed by unconstrained strategic defenses precipitate both offensive countermeasures and efforts to match enemy defenses. Arms control advocates argue that open-ended offensive-defensive arms races will inevitably result from overreliance on technological solutions to security. Technology is neutral. It has no loyalty to offense or defense. Solutions to defending against today's nuclear forces can be negated by tomorrow's offensive breakthroughs.

By contrast, diplomacy and arms control are more cost-effective means to maintain mutual security. Difficulties, however, for reaching long-range arms control agreements can be seen in the long START-SDI deadlock. The Soviets oppose reductions in offensive forces without constraints on strategic defense; the Reagan Administration consistently refused to bargain away SDI development and testing. The issues are more conceptual than technical. Do we want SDI to enhance mutual deterrence or to provide unilateral defense? If the answer is mutual deterrence, arms control may be the most cost-effective option.

Spokesmen for the Reagan Administration described mutual deterrence as an intermediate step on the route to a world of 'impotent', 'obsolete' or 'zero' nuclear missiles. But this defensive utopia is incompatible with mutual deterrence unless both sides can agree on specific levels of offensive and defensive weapons. Arms control solutions have little long-term value unless they can head off competition generated by the offensive-defensive relationship. The first step is to identify the contradictions and the instability created by a defensive doctrine (deterrence) that relies on offensive military strategy. Figure 5.1 illustrates the paradoxes that arms control must

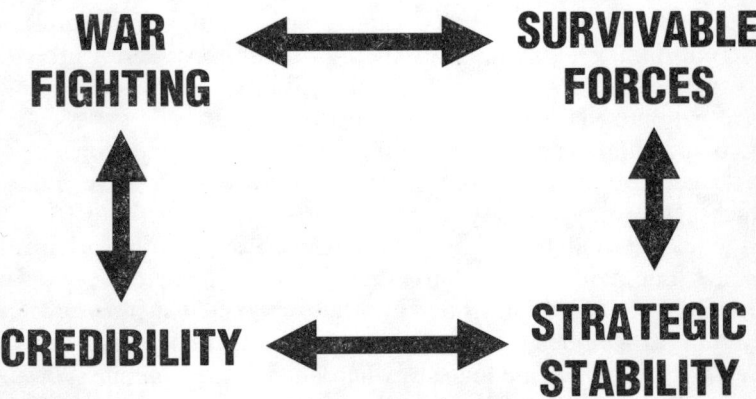

FIGURE 5.1 *Dilemmas of Nuclear Deterrence*

reconcile. Stable deterrence requires survivable nuclear forces that confront the enemy with assured retaliation. Vulnerable forces tempt pre-emption in a crisis. These traditional formulations are widely accepted in the arms control literature and by those who advocate stability through *mutual* deterrence. Translating defensive doctrine into military strategy has, however, produced an entirely different set of objectives in both Moscow and Washington. Deterrence has no credibility if either side fails to deploy forces and develop plans and strategies for their use. The need for credibility drives war plans and strategy. If deterrence fails, current strategy seeks to limit damage to the homeland through preferential attack against enemy military capabilities, especially time-urgent targets or those that are capable of inflicting prompt and massive damage. War fighting strategies of one side that reduce retaliation capabilities of the other negate efforts to deploy survivable forces. Deterrence thus becomes dynamic in the form of competition to achieve offensive advantages while maintaining one's own survivable force structure. The result is an unending cycle of offensive strategy that negates the objectives of mutual deterrence and strategic stability.

Whether nuclear weapons and the contradictions between deterrence and war fighting will continue to dominate Soviet-American relations depends on the scope of future arms control regimes. Deterrence at the end of the 1990s will be determined by offensive modernization programs permitted by the START Treaty, and by technological capabilities advanced by SDI research and development. A significant point often overlooked is the extent of modernization that START allows. Offensive threats that will confront both superpowers in the year 2000 can be seen in the modernization programs in development now. These programs pose serious threats to strategic stability, threats that will need to be addressed by some combination of START II and strategic defense.

Modernization can and should play an important role in achieving the same goals as arms control and strategic defense–deterring an attack. The irony of strategic nuclear doctrine during the Reagan years was the sudden emphasis on strategic defense to render ballistic missiles 'impotent and obsolete', that was combined with a pervasive offensive nuclear modernization program designed to increase nuclear weapons' capabilities for an offensive, hard-target counterforce role. The ability of either side to destroy hardened nuclear missile silos and command posts undermines deterrence stability by creating pressures for pre-emptive strikes during periods of crisis. Before arms control, strategic modernization programs, and strategic defense can be co-ordinated for greater stability, it is important to reconsider the reasons for the ascendancy of destabilizing weapons that weaken deterrence.

Destabilizing weapons are the products of technology and strategy. Both superpowers have been dominated by strategists who believe in active deterrence by denial. This requires war fighting options that pre-empt enemy strategic nuclear weapons in order to limit damage from either his pre-emptive or retaliatory strike. Deterrence by denial is the antithesis of deterrence stability. Two-sided deterrence by denial is mutually threatening, unstable during a crisis, and fuels arms competition in peacetime.

The real question for advocates of pre-emption and deterrence by denial is, does it give you deterrence? It is a fact that both sides view the other as the potential aggressor. Nuclear weapons, political crisis, and benign self-images are volatile combinations. Neither side is likely to go to war if they believe their society would be destroyed as the result of a war they started. But the exception would be a war precipitated by the belief that your territory is about to be subjected to an attack that disarms or destroys. As John Scott has argued, 'the expectation of total destruction is a good deterrent if that destruction is a consequence of one's own initiative, and the worst deterrent if it is the certain and credible intent of another nation'.[2] Ironically, both sides disavow the possibility of winning a nuclear war. Equally, neither side has disavowed the qualitative modernization programs that would allow them some prospect, however desperate, of emerging from a nuclear conflict with some kind of meaningful victory. Both Soviet and American nuclear weapons modernization programs illustrate the problem and the open ended competition fueled by benign self-images.

Such an image is illustrated by the final strategic modernization programs of the Reagan Administration. These systems are designed to support an offensive-dominated world and are inconsistent with the former President's vision of a defense-dominated world. If fully funded they would include:

- 50 additional MX missiles for rail mobile deployment;
- D-5 missile production for Trident submarines 9 and 10 as well as for retrofitting of the first eight Trident submarines in the 1990s;
- new penetration aids for ICBM warheads (advanced strategic missile systems program);

- demonstration tests for maneuverable re-entry vehicles (MARV);
- continued development of the B-2 Stealth or Advanced Technology Bomber;
- development of second generation air-launched cruise missiles and short-range attack missiles (SRAM II) for greater penetration and target flexibility;
- continued deployment of a 500 sea-launched cruise missile force (TLAM-N) aboard 190 ships and submarines;
- upgrades for the B-1B penetration capability;
- continued conversion of all B-52Hs for carrying 20 air-launched cruise missiles each;
- relocatable Target Program to integrate space-based sensors and modernized forces (that is, stealth technology) to attack Soviet mobile ICBMs.[3]

The driving force behind many of these programs was Reagan's pledge to redress the strategic balance that he perceived to be drifting in the Soviet's favor. More recently, strategic modernization programs have been defended in the context of former Secretary of Defense Weinberger's 'competitive strategies' initiative to exploit Soviet weaknesses with US strengths. Both Weinberger and his successor, Frank Carlucci, recognized in their Annual Reports to Congress (FY 1987–90) that strategy and planning needed to be shaped by a more efficient methodology, and resources had to be used more effectively to offset deficit-driven budget constraints. Through its superior technology, the United States plans to 'enhance deterrence by making significant aspects of the Soviet's military capabilities obsolete'. In practice, this means offensive modernization programs that force the Soviets to invest heavily in defensive systems rather than offensive weapons. At some point, competitive strategies are not consistent with administration goals to initiate a transition to a defense-dominated world. Neither are they compatible with an arms control regime designed to support deterrence stability. That point appears to be where a dependence on offensive modernization rather than negotiated constraints drives the offensive-defensive rivalry between the superpowers. If the Bush Administration wishes to encourage effective defense and deterrence stability, it should not encourage competition in hard-target counterforce weapons. Force modernization that is most and least compatible with deterrence stability and a declining reliance on nuclear weapons are contrasted in Table 5.1:

TABLE 5.1   US Offensive Modernization Options

Least supportive of deterrence stability	Most supportive of deterrence stability
1. Additional MX missiles beyond the 50 currently deployed.	1. Modernized Minuteman IIs or Minuteman IIIs in existing silos and /or a rail or roadmobile system.
2. Retrofitting Trident Submarines 1-8 with D-5 ballistic missiles.	2. Retain C-4 SLBM; limit D-5 to one or two Trident Submarines.
3. Mix of BI-B and B2 bombers with advanced technology (stealth) cruise missiles.	3. Retain and upgrade B-1B as a stand-off bomber. Kill the B2 program.
4. Command and control modernization that integrates space-based sensors with missiles and bombers to attack relocatable targets (Soviet mobile ICBMs).	4. Combination of force modernization and arms control to protect mobile ICBMs.
5. Anti-Satellite (ASAT) testing and deployment.	5. Preserve deep space assets against ASATS; develop satellite reconstitution capabilities and Remote-pilotless vehicles (RPVS) to hedge against Soviet ASAT threats.
6. Maintain three strategic bombers with modernized cruise and air to ground missiles.	6. Convert B52G/Hs to conventional missions.
7. Deploy full compliment of programed nuclear sea-launched cruise missiles.	7. Deploy sea and air-launched cruise missiles for non-nuclear theater and strategic missions. Ban nuclear SLCMs.

Whether a START Treaty is stabilizing depends on how the two powers choose to fill their allotments under the agreed ceilings of 1600 launchers, 6000 warheads, and a sub-ceiling of 4900 warheads on ballistic missiles–sea- and land-based. Preferential cuts should be made in weapons most likely to start a war. This means hard-target counterforce weapons which dominate the US strategic modernization program. Table 5.1 suggests potential modernization options that are compatible with strategic stability and the economic realities confronting the defense budget. A top priority is to blunt the cutting edge of a ballistic missile force that is both vulnerable to attack and capable of a pre-emptive first-strike.

Several options are available. Long-term ICBM modernization decisions will be influenced by a combination of cost-effectiveness and survivability. A primary objective for survivability is to increase the ratio of US launchers (Soviet aim points) to Soviet hard-target warheads. The need to prevent a Soviet 'first-strike bonus' can be seen in a scenario in which they launch their SS-18 force against a silo-based MX missile force. Both missiles carry ten warheads. A single Soviet SS-18 could theoretically destroy five silos and their 50 warheads (it is assumed that the Soviets target two warheads per silo). The exchange rate of one Soviet ICBM for five MX missiles (ten Soviet warheads for 50 US warheads) is a dramatic example of deterrence instability created by weapons that invite a first-strike. On the other hand, a single warhead per ICBM would force the attacker to expend two warheads for each US launcher and warhead under attack.

Applying this logic to ICBM modernization under a START regime means that the ICBM force should spread its allotment of warheads over the maximum number of launchers. Partial or total de-MIRVing is required and can be accomplished in a number of ways depending on the nature of Soviet ballistic missile modernization. Under START-mandated reductions of the SS-18 missile from 308 to 154, the United States could defer deployment of a costly mobile ICBM while maintaining a silo-based force of existing single warhead Minuteman IIs and down-loaded Minuteman IIIs (remove two of its three warheads). A force of 900–950 ICBMs in existing silos would be cost effective, requiring some 1800 Soviet warheads for a credible first strike. As a hedge against other Soviet forces (sea-based or modernized SS-24s), a small force of 50-90 MX missiles could be deployed in the Air Force's preferred rail-mobile system. The combination of START reductions and a mixed silo-rail-mobile ICBM force is cost-effective and can be deployed rapidly. ICBM vulnerability could then be assessed against future Soviet programs to determine whether the next phase should be in the direction of further START II cuts, more mobile ICBMs, strategic mobile ground-launched cruise missiles (GLCMs), or a phasing out of the land-based component of the strategic triad. In a perfect world, a single warhead, mobile ICBM would be the preferred option to

ICBM survivability. But too much time has been lost in the decision-development stages of the small, road-mobile ICBM (SICBMs or Midgetman) and its $45 billion cost estimates for 400–500 launchers have left Midgetman with too many political liabilities to be sustained through the procurement battles ahead.

The D-5 or Trident II ballistic missile is scheduled to be deployed with Trident submarine number nine and eventually retrofitted on the first eight Trident submarines during normal overhauls in the 1990s. The D-5 is the first sea-based weapon with a missile silo-destroying capability, making land- and sea-based ballistic missiles interchangeable in their capabilities. It is therefore difficult to justify both the MX and a large Trident II missile force at a time when the Soviets are deploying more survivable mobile missiles and a START Treaty promises to reduce total missile numbers, including a 50 per cent reduction of their most threatening system, the SS-18 heavy ICBM. Initial deployments should be limited to Trident submarines nine and ten with the less destabilizing C-4 missiles remaining on the first eight Trident submarines (saving $940 million through 1994).

Serious consideration should also be given to a new, smaller submarine. START ceilings would limit the US to approximately 18–20 Trident submarines, a cut of more than 50 per cent from previous levels of 41 boats. These reductions make the American sea-based leg of the triad more vulnerable to Soviet advances in anti-submarine warfare. A force of smaller submarines carrying fewer missiles would be a more prudent hedge than investing in a land-mobile system. These will inevitably face limits on their mobility given the political and geographic constraints on deployment areas in the United States compared with the more accessible land mass available in the Soviet Union.

A new submarine would face many of the same cost obstacles as the controversial mobile ICBM. A cost-effective means of achieving the most favorable launcher-warhead ratio allowed by START is to maintain a force of at least 20 submarines–all Trident or a Trident-Poseidon mix. Trident submarines could be downloaded, carrying, for example, 20 missiles rather than 24, or their missiles could carry five warheads rather than eight. Arming missiles with fewer warheads enhances survivability by extending missile ranges and opening larger areas of the oceans for deployment and stationing.

The bomber leg of the strategic triad has been the most costly and has encountered the largest number of technical difficulties in meeting design and mission specifications. Of the $215 billion spent in the past eight years on strategic nuclear weapons, 60 per cent has been invested in bombers.[4] The mix of two new heavy bombers–B1-B and B2–is unaffordable and unnecessary for credible deterrence. The most cost-effective bomber force would be a mix of B-52s and B-1Bs with conversion of the B-1B force to stand-off platforms for modernized (stealth) cruise missiles. The biggest obstacle is Air Force insistence on retaining penetration bombers. Under START and subsequent treaties, however, strategic nuclear forces will become too scarce to target Soviet air defenses and to blast nuclear corridors through the Soviet Union that open primary targets to penetrating bombers. Air-launched cruise missiles can reach primary targets at less cost while still confronting the Soviet Union with major air defense requirements.

The Army's development style for new conventional weapons could be instructive for the Air Force. The Multiple Launch Rocket System (MLRS), for example, is a low-tech launcher for high-tech munitions. The Army builds 'dumb' launchers for 'smart' weapons. The Air Force prefers smart carriers like the B-2, stealth bomber and 'dumb' weapons (gravity bombs and short-range attack missiles). A stand-off mission for strategic bombers would save money and force the Air Force to put its money where it belongs–in weapons, not the delivery systems. Bomber modernization should follow the example of the B-52, the Strategic Air Command's workhorse since the 1960s. The B-52 was designed in 1948 and the models still in service (B-52Gs and Hs) rolled off the assembly line in the late 1960s. Intensive monitoring, extensive rebuilding and updating over the years have made the B-52 responsive to the growing Soviet air defense threat, first with the short-range attack missile (SRAM), penetration aids and electronic counter measures, and more recently with long-range cruise missiles that allow nuclear release far from Soviet territory. The aircraft has undergone 1899 engineering modifications, making it in one supporter's description, 'the aeronautical equivalent of the bionic man'.[5] Despite its age, it ranks consistently below average in Air Force accident ratings. Moreover, the Air Force does not plan to retire its B-52s. They will fly for another decade as conventional weapons carriers. This same life-extension, mission-modification strategy can be applied to the B-1B as an alternative to the more costly B-2 program.

The most serious problem with the elimination of a penetration mission for bombers is the START counting rule conceded by Soviet negotiators. Under the treaty, a bomber loaded with gravity bombs or SRAMs would count as only one launcher and one warhead. The discount against the 6000 warhead ceiling constitutes an incentive for the two sides to retain forces that are better suited for retaliating missions. On the other hand, the strength of Soviet air defense makes discounted penetration bombers less threatening than 2000 advance technology cruise missiles that could be carried by existing B-1B bombers.[6] The B-2 at an estimated $500 million per copy is a cathedral, not a weapon to risk in war (the Air Force wants to build 132). Production has been deferred and the costly program may eventually be killed in favor of continued programs for stealth cruise missiles and the advanced technology tactical aircraft for conventional and theater nuclear missions. An equally important issue, however, is the need for penetration bombers at any price. The rationale for penetration missions should be carefully scrutinized before production decisions are made that may result in a reduced force structure at higher unit costs and greater vulnerability to Soviet defenses.

How to limit nuclear sea-launched cruise missiles (SLCMs) has been a major obstacle at START. Because of their short flight times and accuracy they pose a serious threat to both strategic stability and arms control. The Soviets have insisted that SLCMs be limited or reductions in ballistic missiles would be meaningless.[7] There are three additional reasons why the US should consider a total ban on nuclear SLCMs. First, the nuclear threshold at sea should be as high as possible given the vulnerability of carrier battle groups and battleships to Soviet nuclear weapons at sea. Second, conventional cruise missiles are ideal weapons for low-intensity conflict. These weapons give the President precise and measured military responses against Third World threats without risk to expensive aircraft and pilots. A total ban on nuclear SLCMs would be easier to verify than a mixed conventional/nuclear force. Success in developing a verification regime would protect the option to deploy large numbers of conventional SLCMs. Third, the United States with an exposed coast line is far more vulnerable to a sea-launched cruise missile attack than the Soviet Union.

The retention of nuclear SLCMs could be more attractive in the context of efforts to increase the survivability of sea-based deterrent forces. For example, a force mix of Trident submarines and SLCMs on other classes of submarines would contribute to stability by creating a greater diversity and dispersion of START accountable forces. This option should, however, be evaluated against the total benefits of a START Treaty and the risk that the SLCM issue could block a final agreement.

Finally, SDI must be brought into the arms control-strategic modernization equation. Deterrence and fiscal stability require that the SDI bargaining chip be cashed at the negotiating table. The concession need not be in the form of abandoning SDI. Limiting development and testing to the strict interpretation of the ABM Treaty for a ten-year period or more in exchange for a 50 per cent reduction in Soviet nuclear weapons is a small price to pay. The interim period gives both sides the opportunity to think through the most mutually beneficial mix of offense and defense.

Limiting offensive-defensive competition requires mutually balanced constraints. Ascendancy on one side of the equation by either party is certain to provoke short-term countermeasures on the other side and long-term competition in both offensive and defensive technologies. Negotiating leverage and willingness to accept constraints in both offense and defense are two sides of the same coin. Failure to compromise will risk Soviet efforts to match US defenses and counter its offensive modernization programs. Mutual fear that the other side may develop offensive means to penetrate defenses will fuel competition in both. These are the strategic obstacles to reaching President Reagan's dream of offensive reductions in a defense-dominated world. Arms control that includes both defensive and offensive constraints, and unilateral basing and deployment solutions to the problems of weapons vulnerability, offers the most cost-effective and stable solutions to the problems confronting strategic modernization.

This, in essence, was the conclusion of the Scowcroft Commission Report endorsed by Reagan on 29 April 1983. The bipartisan report recommended the deployment of a small, mobile ICBM, smaller and more numerous submarines, improved communications, and increased penetration effectiveness of retaliatory forces. These steps in some combination are more likely to lead to successful arms control negotiations and strategic stability than the offensive-defensive arms competition that is likely to grow out of the Reagan legacy of combining destabilizing offensive modernization programs and the Strategic Defense Initiative.

The proponents of SDI and counterforce modernization programs risk falling into the trap described by Freeman Dyson as the 'technical-follies future'. The 'technical-follies' is the open-ended offensive-defensive arms race that will be precipitated by unilateralists who advocate deployment of strategic defenses even in the absence of arms control.[8] The unilateralists who attempt to create a security regime based solely on technology can never rest. Soviet responses, sooner or later, negate the best efforts of those who abandon diplomacy in their search for security through technology. Equally important is what sacrifices the Soviets are willing to make in their own strategic modernization programs.

SOVIET STRATEGIC MODERNIZATION AFTER START

The evolution of Soviet strategic doctrine described in Chapter 2 has reached a new junction under the leadership of Mikhail Gorbachev. His new military thinking embraces arms control, reductions in Soviet defense spending, and 'non-offensive defense'. For strategic nuclear weapons this means maintaining survivable weapons capable of second-strike, retaliatory missions.

Has doctrinal change been matched by significant shifts in Soviet force structure as projected in current strategic modernization programs? The direction is by no means clear, although trends point to a more survivable force structure that is more evenly distributed across the three legs of the Soviet triad of land- and sea-based ballistic missiles and long-range bombers.

The Soviets are deploying two new ICBMs–a road-mobile, single warhead SS-25, and a ten-warhead SS-24 that is rail-mobile. Both missiles are capable of maintaining a higher state of readiness and are more accurate than the silo-based SS-11s and SS-17s that they are replacing. Neither missile, however, has the first-strike potential of the SS-18. Nevertheless, in the aggregate, a Soviet ICBM force comprised of SS-25s, SS-24s and a smaller, but also modernized SS-18 force could improve Soviet capacity to destroy hardened targets. Unless constrained by START and unilateral US deployment decisions, Soviet ICBM modernization solves only half the deterrent stability problem – more survivable Soviet ICBMs. There are, however, compelling reasons why the Soviets may be willing to deploy these forces in numbers that significantly reduce the threat to US retaliatory forces. The combination of threats to Soviet silo-based forces (MX and D-5 ballistic missiles) and the Strategic Defense Initiative have forced the Soviets to deploy a less ICBM-dominant force structure. Under START limits, a larger percentage of Soviet strategic forces will have to be deployed on submarines and aircraft.

Hedges against SDI and offensive weapons are certain to continue even if START sets significant short-term restrictions on both offense and defense. This gives American negotiators an opportunity to push for further reductions in the SS-18 force and to ban or severely limit deployment of MIRVed mobile ICBMs. With warhead limits of 6000, the Soviets are unlikely to feel secure with 25 per cent of their START allowable warheads deployed on SS-18s which comprise only ten per cent of their launcher limits. This unfavorable US warhead to Soviet target ratio would be a foolish sacrifice of survivability for lethality. Since the same is true of the MX deployment decision, a mutual solution to minimize the hard-target first-strike threat from ICBMs is a lower, but equal limit of 50–100 MX and SS-18 ICBMs. This would open the door for the administration to defer a new mobile ICBM force in favor of a less expensive 1000 silo-rail mobile force described above. Further US ICBM modernization could then await the agenda for START II, a treaty that should be aimed at refining deterrence stability, and less on dramatic new reductions. At that time, a decision to build a mobile missile could be evaluated as a compromise with the total elimination of land-based missiles in favor of more submarines and bombers.

Soviet submarine and bomber force modernization are, for the time being, less destabilizing than their ICBM programs. Two new Soviet submarines, the Typhoon and the Delta IV, armed with SS-N-20 and SS-N-23 missiles and their follow-ons will enhance the survivability of Soviet submarines by enabling them to attack targets in the United States while remaining close to home waters. Improvements in survivability have not been matched by missile lethality and accuracy to match the D-5's hard-target capability. Soviet missiles currently under development do not approach even the Trident I (C-4 missile) in accuracy and yield.[9] An active research and development program can be assumed to eventually match US capabilities, but the combined effects of American anti-submarine warfare (ASW), geographic constraints on Soviet submarine deployments, and arms control constraints give American planners time to re-evaluate ICBM vulnerability after the turn of the century when Soviet submarines may put land-based ballistic missiles at risk.

The new Soviet long-range bomber, the Blackjack, is entering service and, with the older design Bear H bomber, serves as the Soviet's air-launch cruise missile force. The CIA estimates that Soviet cruise missile forces could grow to more than 2000.[10] This number is contingent on the number of bombers actually deployed. The Soviets have not placed great faith in the manned bomber leg of their strategic triad. It is conceivable, however, that a force of 100 or more bombers could provide a hedge against ballistic missile vulnerability while forcing the US to spend large sums to upgrade its air defense system. The Soviets too can play competitive strategies. On the other hand, a modernized bomber force would be consistent with the second-strike strategy proclaimed in Gorbachev's new military thinking. Bombers are much less suited to pre-emptive, damage limiting strategies associated with ballistic missiles.

The key to Soviet force modernization and final compromises at START will be the Bush Administration's position on strategic defense. Soviet officials reject descriptions of SDI as 'defensive'. Gorbachev and others maintain that 'space strike weapons' are offensive in character and could be used to control all access to space as well as to menace targets on the ground and in the atmosphere.[11] More threatening to Soviet military doctrine and strategy is their long-standing view that defense is not an end in itself, but parallels developments in offensive forces. The real purpose ascribed to SDI by the Soviets is, in combination with offensive force modernization, to achieve a first-strike capability against the Soviet Union:

> What we are dealing with in reality are measures that are part of an overall offensive plan directed at upsetting strategic parity, achieving military superiority, and preparations for delivering a nuclear first-strike.[12]

Soviet attacks on the Reagan drive for a defense-dominated world have obvious propaganda value in support of Soviet arms control objectives, but this does not mean that Soviet assessments are entirely off the mark. American critics have pointed to the offensive modernization programs described here that are inconsistent with a transition to strategic defense.

Soviet efforts in their own amply funded strategic defense research program are seen by some as evidence of Soviet duplicity–propaganda aimed at killing SDI while massive investments in their own ballistic missile defense system continue. Both sides have maintained sizeable research and development efforts in missile defense since the 1972 ABM Treaty. Breakthroughs in new technologies (laser and particle-beam research) during this period led to President Reagan's optimistic 'star wars' speech in 1983, and the consolidation of all ballistic missile defense research under a new Strategic Defense Initiative Office (SDIO) in 1984.[13]

The Soviet SDI is much like the pre-1983 American effort. It is characterized by compartmented pockets of research and development that lack a centralized plan or systemic architecture of the kind that characterizes the Reagan concept of space-based, layered defenses controlled by complex, high-speed computer battle-management systems. If Soviet programs were on the verge of significant breakthroughs, they would not have become the centerpiece of Gorbachev's arms control strategy. These sacrificial lambs are consistently offered in exchange for offensive reductions and adherence to the ABM Treaty.[14]

SHAPING STRATEGY AND TARGETING POLICIES FOR DETERRENCE STABILITY

Nuclear weapons' modernization programs reflect and are shaped by military strategy and target priorities. Deterrence stability cannot be achieved without significant changes in both. Change is possible because the elements of a stable deterrent strategy are in place. Only targeting priorities need to be modified in a way that separates first-strike incentives from less volatile war plans that promote *mutual* deterrence. This problem and its evolution requires a brief review of nuclear strategy.[15]

Atomic Monopoly, 1945–49

Strategy and targeting policies evolve gradually. They are shaped by the numbers and capabilities of nuclear weapons in our stockpile, the range and accuracy of our delivery vehicles, and by strategic intelligence that allows American military planners to find targets in the Soviet Union and assess their importance in deterring or defeating the Soviet Union in a general war.

Nuclear targeting during the era of 'atomic monopoly' seems primitive by the standards of the 1990s. The few atomic bombs (50 in 1948) in our arsenal were large, awkward to handle, and carried by a limited number of aircraft flown by a small number of qualified crew members in airplanes that required European bases for refueling. Their targets were also 'simple'. Given limited intelligence on military targets in the Soviet Union, and with a few weapons available, we planned to attack what we could find–cities and industrial complexes.

The primary targets during this period were industrial. Soviet aggression would be deterred or defeated by destroying the capacity and will to wage war. Population *per se* was not targeted, but since workers and industries are inseparable, large numbers of casualties that would have resulted from industrial attacks represented in General Curtis LeMay's words, a good 'bonus effect'. At no time during this period was the continental United States at risk from a Soviet attack. The American deterrent was aimed solely at Soviet aggression in Europe.

The Road to Massive Retaliation, 1950–60

Planning during the Truman Administration's final years grew more complex as a result of the growing nuclear stockpile and the expansion of the Strategic Air Command (SAC) to deliver it against Soviet, East European, and eventually, Chinese targets. The emergence of Soviet atomic forces and the formalized American commitment to defend Europe (both in 1949) resulted in vast increases in the nuclear stockpile (450 weapons by 1950 and 1000 two years later) and more complex war fighting plans. By the early 1950s, three distinct types of targets appeared in US war plans. The code names *Bravo*, *Romeo*, and *Delta* signified that objectives in a nuclear war were the following: (1) blunt Soviet capabilities to produce or deliver atomic bombs; (2) retard the movement of Soviet conventional forces in Western Europe through attacks on Soviet transportation and communications; (3) disrupt the Soviet capabilities to support war through attacks on specific industries such as electric power, petroleum, and atomic industry.

The Strategic Air Command's basic war plan in March 1954 was to launch 735 bombers in a single, massive attack against the Soviet Union, Eastern Europe, and China. President Eisenhower planned to deploy tactical nuclear weapons to meet limited wars in Asia and Western Europe, but in a general war against the Sovier Union there was resistance from the Air Force to the concept of limited war. Neither was there the slightest distinction or even awareness of political divisions that separated the Soviet Union from its East European allies or from some of its own non-Russian republics. Military exigencies alone drove American nuclear war planning. These plans by 1954 fully reflected both a capacity and intention to execute

Secretary of State John Foster Dulles' proclamation of 'massive retaliation' (January 1954) as the preferred strategy for dealing with Soviet aggression.

The latter years of the Eisenhower Administration saw growing resistance to the Strategic Air Command's domination of nuclear strategy and planning. Pressures from many sources for a more 'flexible' nuclear response would soon overwhelm SAC's resistance to the concept of limited nuclear options. One of the first steps in the process was President Eisenhower's creation of a joint military staff (Joint Strategic Target Planning Staff) charged with preparing a national strategic target list. This list became the basis for the first fully integrated nuclear war fighting plans, the Single Integrated Operational Plan or SIOP. From Eisenhower to Bush, the SIOP has been the basic employment plan for US nuclear weapons.

The Road to 'Limited' Nuclear War, 1961–90

The development of more 'flexible' nuclear war fighting plans was left to the Kennedy-McNamara defense planners who had both the tools and the inclination to change nuclear strategy in several dramatic, but often confusing ways. One of the most dramatic changes in capabilities came in the form of strategic intelligence. American U-2 flights over the Soviet Union from 1956 to 1960, and the development in the early 1960s of satellite photography gave nuclear targeteers precise information on the location of Soviet targets. This capability gave planners the ability to develop war plans that were more discriminating (flexible) between military and civilian targets. Consequently, greater emphasis was placed on targeting both Soviet conventional and nuclear war fighting capabilities.[16]

Secretary of Defense McNamara was reportedly struck by the exuberance of US nuclear war plans. One of his first steps in developing the Kennedy Administration's SIOP was to move away from massive retaliation and toward more flexible targeting options. He first separated Eastern Europe and China from initial attacks on the Soviet Union. Strategic reserves were planned, and as nuclear forces became more survivable from a Soviet attack (nuclear missiles in submarines and ICBMs in hardened silos), declaratory policies emphasized that American strategy was to launch a second, but more limited nuclear strike in response to a Soviet attack on the United States or Western Europe.

There were two elements of this strategy that caused some confusion among defense intellectuals who watched these developments from outside the administration. First, McNamara introduced the concept of 'damage limitation'. This required attacks on Soviet military targets before they could be committed against the United States or Europe. The second element of the strategy was 'assured destruction'. This required attacks against Soviet cities and economic targets to punish (therefore deter) the Soviets for attacks on the US or its allies in Europe or Japan. The two elements of the strategy represented 'rungs' on the ladder of escalation. Military targets would presumably be struck first, and cities only if war escalated to an all-out nuclear exchange.

McNamara's strategic flexibility ran into several difficulties. At the political level, critics attacked his emphasis on war fighting (attacking military targets). A second-strike capability against military targets could not be distinguished from a first-strike capability and would, therefore, be destabilizing to deterrence. At the military level, the emphasis on military targets gave the Air Force and Navy a virtually open-ended target list to justify increases in the nuclear stockpile and delivery systems. To extract himself from the contending pressures of outside critics and bureaucracies all too eager to support and expand nuclear war fighting capabilities, McNamara appreared to reverse himself. By 1967 he declared 'assured destruciton'–the ability to destroy an attacker as a viable twentieth-century nation–is what provides deterrence, not our ability to partially limit damage to ourselves.

The shift in declaratory policy gave the false impression that war plans were aimed only at Soviet cities and industry. In fact, war plans did not change significantly. Extensive coverage of military targets still remained in the SIOP. McNamara publicly described only the most extreme option. In doing so, he was establishing a procurement strategy rather than a war-fighting strategy. The criteria for assured destruction were used by McNamara as a measure for deciding force levels and denying the military services and Congress funds for additional strategic forces.

McNamara's emphasis on deterrence through the assured destruction strategy set the stage for what appeared to be dramatic changes in nuclear strategy in the 1970s. Beginning with Secretary of Defense James Schlesinger, declaratory policies began moving away from an emphasis on assured destruction to more limited nuclear war-fighting options. The President, it was argued, should never be left with the single option of ordering mass destruction of enemy cities in the face of certain retaliation against American cities. Such a strategy lacked both morality and credibility. The President needed options. If deterrence failed, he must mitigate the consequences of nuclear war by a capability to attack all categories of military targets, control escalation before cities were attacked, bargain during a nuclear exchange, and if possible, terminate nuclear conflicts on terms 'favorable' to the United States at the lowest level of destruction.

The first official public discussions of these issues came in President Nixon's foreign-policy message to Congress on 18 February 1970:

> Should a President, in the event of a nuclear attack, be left with the single option of ordering the mass destruction of enemy civilians, in the face of the certainty that it would be followed by the mass slaughter of Americans? Should the concept of assured destruction be narrowly defined and should it be the only measure of our ability to deter the variety of threats we may face?[17]

A series of studies and directives followed, providing political guidance on structuring more flexible pre-planned nuclear response in the US war plan. Secretary of Defense James Schlesinger publicly announced the change in targeting strategy. Assured destruction and the old policy of initiating a suicidal strike against the cities of the other side 'were no longer adequate for deterrence'. He would, therefore, 'implement a set of selective options against different sets of targets on a much more limited and flexible scale'.[18] Military targets were given top priority. By adopting the strategy of limited nuclear options, planners reasoned, escalation might be averted short of attacking target categories in major urban-industrial centers.

The Carter Administration refined the limited nuclear war strategy by de-emphasizing Soviet economic targets (moving still farther away from mutual assured destruction) and stressing the importance of survivable American strategic forces and command and control systems required to execute a limited nuclear war.[19] Subsequently, the Reagan Administration produced a Nuclear Weapons Employment and Acquisition Master Plan, which maintained the legacy of limited nuclear warfare and stressed the requirements for strategic modernization.

Under McNamara, the false impression was created that war plans were still governed by a single, massive strike and that the nuclear 'wargasm' of the 1950s still represented American strategy. Options did exist. What changed most after 1973 was the realignment of declaratory and employment policies. Prior to 1973, the two had been dramatically at variance, with the result that nuclear targeting strategy had been grossly misrepresented.

The move toward more flexible options in the 1970s was evolutionary and less dramatic than public debate would indicate. It is true that important changes in the strategic landscape were pushing this evolution in new directions. The Soviet Union achieved strategic parity. Both sides were developing more accurate ICBMs which made it possible to attack missiles in their silos and hardened communications and control facilities. These developments accelerated the evolution of limited nuclear options in the Carter and Reagan Administrations. (See Table 5.2 for the Evolution of US Strategic Doctrine). If Reagan's policies stand out, it is only because he, for better or worse, attempted to fund and deploy the strategic systems required to fully execute a 'flexible' nuclear strategy that

TABLE 5.2 *Strategic evolution*

	Declaratory Policy	Employment Plans
Truman	Assured destruction	– limited capabilities – attack with atomic and conventional bombs
Eisenhower	massive retaliation	– massive attacks against cities and military targets throughout communist bloc
Kennedy	flexible response, emphasis on assured destruction	– assured destruction – damage limitation – massive attacks on cities and military targets – geographic withholds
Nixon-Reagan	limited nuclear options	– packaged nuclear responses – 'small' to large – priority to damage limitation – escalation to assured destruction

incorporated the old McNamara elements of 'damage limitation' (counterforce targets) and 'assured destruction' (escalation to countervalue targets).[20]

The evolution of limited nuclear war options since 1973 has been both good and bad. Its most laudatory aspect is the attempt to mitigate the consequences of total nuclear war through limited attacks if deterrence fails. This is offset by the requirements for pre-emptive strikes which make nuclear war more likely. The delicate balance between mitigating the consequences of war and increasing the risk of war can be stabilized through a combination of arms control and modernization programs that reduce hard-target, counterforce capabilities and, in turn, force a shift in targeting priorities.

Mutual deterrence has been replaced by a strategy of mutual engagement at the expense of crisis control. To paraphrase Ashton Carter, recent administrations have been 'polishing a hair trigger' when they should have been building better safeties and improving the system's eyes (command, control, and intelligence capabilities).[21] The assumptions of rationality and control which underly nuclear deterrence take us only to the edge of crisis. At that point, stress, chance, uncertainty, and fallible men and machines on both sides take over. This is why incentives for striking first must be minimized.

Deterrence stability does not require a return to the city-busting mutual assured destruction (MAD) strategies of the 1950s. Stability requires moving away from the elaborate scenarios of the nuclear war fighters who plan attacks against specific military and leadership targets in a complex nuclear exchange that is hypothetically limited and managed over an extended period. The hard-target counterforce approach to nuclear war fighting does not eliminate the essential assured destruction character of nuclear war for several reasons: first, only in theory is it possible surgically to attack counterforce targets (nuclear and conventional military forces and their controlling 'leadership' targets) while holding countervalue targets (urban-industrial centers, transportation, energy) hostage to the Hobsion choice of escalation or surrender. In reality, the close proximity of countervalue and counterforce targets, the destructive power of nuclear weapons, and collateral damage to people and property would create perceptual problems among the political leadership that is under attack. In a crisis, they would have difficulty distinguishing between a limited attack and a phased, general nuclear war. This point was made more than 25 years ago by chairman of the Joint

Chiefs of Staff, General Lyman L. Lemnitzer, in his briefing of existing nuclear war plans to President Kennedy:

> There is considerable question that the Soviets would be able to distinguish between a total attack and an attack on military targets only....Because of fallout from attack of military targets and co-location of many military targets with (cities), the casualties would be many million in number. Thus, limiting attack to military targets has little practical meaning as a humanitarian measure.[22]

Subsequent studies and computer simulations support General Lemnitzer's fears. One recent study of casualties in a 'limited' counterforce exchange against nuclear weapons and their supporting infrastructures estimated that the *direct* effects of blast, fire, and radioactive fallout could kill between 12 and 27 million people in the United States and a comparable number (between 15 and 32 million) in the Soviet Union.[23]

More fundamentally, it does not seem conceivable that escalation to general nuclear war would not occur given the fragile dynamics of threatening either side's retaliatory forces with a 'pre-empt or be destroyed' choice. This is especially true of Soviet military strategy described in Chapter 2. No endorsement of limited nuclear options or the desirability of limited nuclear dueling can be found in Soviet concepts of intercontinental nuclear war. Moreover, neither power has developed rules, strategies, or mechanisms that connect the threat of escalation to war termination. We declare it, therefore it must exist. Yet for the past two decades both superpowers have deployed counterforce weapons while ignoring the probability that large-scale application of nuclear weapons against the other's strategic forces will not be qualitatively different from their application to countervalue targets. As the study quoted above concluded, 'In view of the massive civilian casualties counterforce attacks would entail, threatening to execute such attacks can be no more credible than threatening to destroy cities'.[24]

In a nuclear war, we are fated to live in a world of mutual assured destruction. The search for damage limiting strategies may be futile given the power of nuclear weapons, the size of nuclear stockpiles even after START, the collateral damage associated with the use of nuclear weapons against military targets, the absence of strategic area defense, the robust size of a 'limited' attack required to destroy strategic nuclear systems, and the uncertainties of controlling escalation.

An alternative strategy can be built on the most stable elements of MAD and limited nuclear war fighting. Massive attacks on urban-industrial targets can be de-emphasized in favor of limited nuclear options and implicit threats of escalation that begin with truly limited (compared with large-scale attacks required of deterrence by denial strategies) attacks on critical industrial, energy, transportation, or conventional military targets. Targeting policies should reorder existing priorities in a manner consistent with deterrence stability, namely, limited options in support of deterrence by *punishment* instead of hard-target counterforce targeting in support of deterrence by *denial*. A more stable mix of countervalue and counterforce targets is the only strategy compatible with mutual deterrence. The threat of limited attack by forces that have a vastly diminished capability for pre-emptive and disarming first strikes will strengthen the prospects for crisis control, intrawar deterrence, and conflict termination.

The United States can accomplish these goals by designing the elements of its nuclear strategy and its force structure in ways that maximize their survivability and minimize their first-strike potential. Failure to do so makes the irrational rational and the unthinkable a requirement for survival in a future Soviet-American crisis. Table 5.3 illustrates the broad choices available with and without the regulatory mechanism of arms control. The matrix contrasts the most and least stable postures in which both lethality and survivability of weapons are in competition. Low survivability against high lethality is the least stable posture (Cell B); mutually low lethality against high survivability is the most stable (Cell C). The dynamics of the arms race, perceived threats, and military strategy drive both sides to deploy forces that are high in both variables (Cell A). This posture can also be stable when competition is regulated by an arms control regime. Table 5.4 applies this logic to specific weapons systems, and illustrates a destabilizing trend in Soviet-American modernization programs that favors considerations of lethality over survivability. The source of greatest instability is fixed, land-based missiles that combine high lethality and low survivability (MX and SS-18). Arms control agreements that reduce lethality, and unilateral modernization programs that enhance survivability are essential prerequisites to the strategic reforms described above.

TABLE 5.3 *Stable and Unstable Strategic Postures**

	Survivability of strategic forces under attack	
	HIGH	LOW
High lethality of strategic forces in counterforce missions	A. Deterrence stability if regulated by arms control regime	B. unstable deterrence; defender driven to first-strike and/or modernization
Low	C. High deterrence stability	D. Strategic uncertainty; unsatisfactory for either side

* CELL C = most desirable posture
  CELL B = most undesirable posture
  CELL A = most likely posture via arms control

The competition between survivability and lethality illustrated in Table 5.4 represents a larger competition in arms control between the political objectives of arms race and crisis stability and the military requirements to execute operational plans and procedures. Judgments about military sufficiency tend to favor the status quo. The search for crisis and arms race stability should not be held hostage to those whose primary interests are in protecting current operational capabilities. Deep reductions require, at some point, new forces, new strategies, and new targeting philosophies that breathe credibility into strategic forces, but not at the cost of stable, mutual deterrence.

TABLE 5.4 *Characteristics of U.S. and Soviet Strategic Forces*

	Lethality*	Survivability*
Fixed ICBMs		
Minuteman III	moderate	low
SS-18	high	low
SS-19	high	low
MX	very high	low
Mobile ICBMs		
SS-24	high	moderate
SS-25	high	moderate
Midgetman	very high	moderate
MX	very high	moderate
Cruise Missiles		
ALCM	very high	moderate
SLCM	very high	high
Bombers		
Stealth	high	moderate
Blackjack	high	moderate
SLBMs		
Trident I	low	high
SS-N-20	low	moderate to high
Trident II	very high	high

* *Lethality* is a combination of accuracy, yield, and time or stealth to target
Survivability includes invulnerability to attack and ability to penetrate enemy defenses
Source: Supporting data compiled by Union of Concerned Scientists

Nuclear weapons, in all likelihood, are with us forever. But arms reductions and a stable, well managed strategic relationship with the Soviet Union are possible, and are as close to a post-nuclear world as we are likely to come. Stability at the most threatening end of the conflict spectrum is possible. But nuclear stability can not be isolated from the threat of conventional war in Europe, the most likely catalyst to an intercontinental nuclear war.

STRATEGIC STABILITY AND CONVENTIONAL DETERRENCE

Chaper 1 describes the linkage between conventional and nuclear forces and the delicate mechanism that controls escalation from one level of warfare to the other. Strategies and force structures that successfully de-emphasize nuclear weapons and threats of escalation depend on credible conventional deterrence. The post-nuclear world defined here is anchored by strong conventional forces. The credibility of NATO's conventional forces in the future will be determined by a combination of modernization programs and the outcome of the Conventional Forces in Europe Negotiations (CFE) which began in March 1989. In spite of Gorbachev's unilateral reductions which preceded the formal negotiations and the general conciliatory tone of Soviet diplomacy, difficult negotiations lay ahead in Vienna for the 23 members of NATO and the Warsaw Pact assembled to test the Soviet commitment to 'new military thinking'.

As American negotiators undertake conventional arms control with their NATO allies, two 'centers of gravity', one political and one military, will be critical to the negotiations. The political center of gravity is the cohesion of the NATO Alliance. This has been a primary target of Soviet diplomacy. Arms control and conventional modernization decisions must be made within the broader objective of maintaining alliance cohesion. Without a united Western front, there is no possibility for credible conventional deterrence in Europe.

NATO's political center of gravity is the foundation on which the alliance has fielded military power sufficient to threaten the Soviet military center of gravity in Europe. That decisive point is the ability of the Soviet Army to maintain offensive momentum on the battlefield. War or political intimidation as a means to Soviet political objectives requires the threat of surprise attack and rapid military victory. Protracted conflict or stalemate on the battlefield poses serious threats to the cohesion of the Warsaw Pact. Unreliable allies may begin to question the cost-benefits of war, just as the Rumanians did during World War II. Their divisions fought with the Germans as far as Stalingrad. But when the fortunes of war turned and the Red Army reached Rumanian soil, they joined with it to crush the Nazi's. Similarly, in a stalemate, Soviet leaders have good reason to fear that national strategies for survival among their East European allies would prevail over Soviet political objectives.

There are other risks. Long and vulnerable supply lines between the West European front and Soviet industrial centers would be difficult to maintain at levels required to meet the rapacious logistical appetites of mechanized divisions and their supporting firepower. There are also the risks discussed in Chapter 3 that protracted war as a challenge to central authority may set off the centrifugal forces of nationalism among Soviet minorities, especially those in the politically strategic Union Republics contiguous to East Europe. These are the intertwined political-military dimensions of strategy that contribute to Soviet self-deterrence if confronted by credible NATO conventional defenses.

Conventional arms control and modernization programs can shape a strategic environment that further degrades Soviet capacity for momentum and quick military victory. The inherent advantage of the attacker in gaining the initiative over the defender must be reversed before Western interests are secure. The growing lethality of NATO's conventional forces and Gorbachev's new military thinking in the form of non-offensive defense make this possible for the first time in post-war Europe.

NATO's broad objective is to achieve stable deterrence by denying Warsaw Pact capabilities for short-warning attack, and the embodiment of that threat in Soviet armored divisions and artillery. These, Philip Karber has argued, are 'the root of military instability in Europe'.[25]

This broad objective can be pursued through a two-front arms control strategy; one to reduce offensive structure and a second to restrict operational capabilities. Structure and capabilities are distinct components of conventional forces. They are the critical variables of conventional arms control. Operational capabilities are the activities of military forces in the field and include training exercises and troop concentrations that can be observed and monitored. Under the provisions of confidence and security building measures (CSBM), on-site observations of training exercises are already in practice as the result of the Conference on Disarmament in Europe (CDE).[26] Under its provisions, the exchange of military observers provides the framework for an expanded conventional arms control verification regime. Supported by national technical means for monitoring Soviet troop movements, on-site observers promote the transparency of Warsaw Pact territory that is required to decrease the probability of a successful attack. Confidence building measures require equal progress in reductions and modifications of Soviet forces in Europe. Several Western negotiating strategies have been proposed:

- disproportionate reductions in primary weapon systems where one side has a numerical advantage;
- equal percentage reductions of total forces;
- reductions in non-equivalent systems (for example, Soviet tanks for NATO aircraft);
- creation of weapons-free zones or partially demilitarized zones;
- redeployment of forces.[27]

The immediate obstacle common to all negotiating strategies is disagreement over the data base from which negotiators begin their efforts to reshape the military balance in Europe. On the eve of the negotiations, *Pravda* published the Soviet Union's most detailed estimates of the conventional balance in Europe. Soviet data reinforces their claim that a rough parity exists between East and West.[28] Discrepancies between NATO and Warsaw Pact data are explained by different weapons aggregations and counting rules which threaten to deadlock negotiations if either side insists on a narrow bean counting approach. Several key examples can be cited and are summarized in the box on p. 162.

There is virtually no prospect for conventional arms control if negotiations become mired in disputes and mutual recriminations over the military balance. A treaty does not require meticulous calibration of opposing forces to achieve mutual security. Domestic political factions and public opinion may be reassured by the appearance of balance and equality, but no historical data exists to support a relationship between military parity and the absence of aggression.[29] Other factors are more important in achieving credible conventional deterrence against the primary Soviet center of gravity in Europe–surprise attack and momentum on the battlefield culminating in a quick victory.

This chapter makes no attempt to summarize the burgeoning literature on conventional arms control. There are, however, two issues critical to this book's central thesis that trends are emerging, and if willfully pursued, can move the superpower rivalry toward a post-nuclear world, a world more hospitable to conventional deterrence and defense. The issues are: (1) what is conventional stability?; and (2) how should it be linked to nuclear weapons and NATO's strategy of flexible response?

In the broadest sense, conventional stability like deterrence in general is a political-military posture that preserves NATO's political

> **THE CONFLICTING DATA BASE FOR NATO AND WARSAW PACT CONVENTIONAL FORCES**
>
> *Ground Forces.* Soviet figures claim rough parity with 3.5 million Warsaw Pact soldiers facing 3.6 million NATO troops. Soviet data includes naval forces, but exclude most support or construction units. NATO excludes naval forces, but count most Soviet construction troops and claim a Warsaw Pact advantage of 3.1 million to 2.2 million troops.
>
> *Tanks.* Soviet data concedes a Warsaw Pact advantage of 2-1 in total numbers of tanks (59 470 to 30 699); NATO claims a 3-1 Soviet advantage (51 500 to 16 424). The Soviets count all tanks – heavy, light, light-amphibious. NATO figures include only heavy, main battle tanks.
>
> *Artillery.* NATO figures include only heavy artillery (100mm and over). Soviet forces have these weapons in great abundance to support ground forces. By contrast, NATO has far fewer of these weapons, but large numbers of smaller (below 100mm) weapons that are organic to its ground forces. Soviet data includes all artillery regardless of caliber (down to 75mm artillery and 50mm mortars).
>
> *Combat Aircraft.* The Soviets insist that NATO has a 1.5-1 advantage in front-line ground attack aircraft. This contrasts with NATO estimates of a 2-1 Warsaw Pack advantage. The large discrepancy is explained by Soviet inclusion of NATO's naval aviation capable of fighting from carrier battle groups in the European theater, and by Soviet definitions of 'offensive' aircraft as ground attack and fighter interceptors as 'defensive'. Soviet definitions ignore multirole aircraft, exclude Soviet medium-range bombers, and oversimplify the offensive-defensive capabilities of tactical aircraft.

cohesion while threatening the Soviet military center of gravity in Europe. This requires careful co-ordination of arms control mandated reductions and modernization of conventional forces that will remain to deter war in Central Europe.

Gorbachev's incentive for arms control can be seen in the sheer size of his army. As the largest conventional force in the world, it is both militarily impressive and economically stifling. The investment required to maintain and modernize it at current levels makes it impossible for Gorbachev to execute his economic restructuring and domestic reforms. The scope of the problem can be seen in the diversion of resources since the Khrushchev era. At the time of his removal, he bequeathed Brezhnev a force structure of approximately three million men supported by 35 000 tanks, and with 26 divisions deployed on foreign soil. Two decades later Gorbachev inherited a military force of 5 500 000 men supported by over 50 000 tanks, and with 40 divisions stationed outside Soviet territory.[30]

There is considerable justification for disproportionate reductions on the Soviet side. Senator Sam Nunn, Democratic chairman of the Senate Armed Services Committee, proposed a strategy that has the appearance of political equality, but produces disproportionate reductions in Soviet military forces. Nunn favors a 50 per cent reduction of the forward deployed forces of both superpowers (two-plus US divisions from West Germany for 13-plus Soviet divisions from East Europe). Withdrawn forces would be redeployed to locations that require equal time to return to their forward positions, thus compensating the United States for its geographic disadvantages.[31]

Senator Nunn's proposal highlights the importance of geography to the negotiations. The vast region involved in negotiating reductions from the Atlantic to the Ural Mountains and the Soviet advantage of proximity create challenges that cannot be solved by disproportionate force reductions. One approach is to divide the region by subzones that are each addressed by specific arms control requirements and by unique NATO force modernization requirements. The NATO plan and the prestigious and often prescient Soviet Academy of Sciences have both proposed to divide the ATTU region into three zones (see Map 5.1: (1) the Central Front; (2) a 'middle or reinforcing zone'; and (3) an external or reserve zone.[32] For each zone 'parity' is defined as percentage reductions, much like the Nunn proposal that places the greatest burden on the side with superiority in a given category of weapons. Significantly, aircraft are included only in the total ATTU region because their range and flexibility do not facilitate constraints in narrow geographic areas.

The Nunn proposal and its unofficial Soviet counterpart preceded formal negotiations and are more ambitious than the opening NATO position in Vienna. Western negotiators seek parity at 10 per cent below NATO levels in the most offensive weapons–tanks, artillery, and armored personnel carriers. Neither the United States nor the Soviet Union would be permitted to deploy more than 30 per cent of these totals (3200 tanks and 1700 artillery pieces) in any one allied country.

MAP 5.1 *Conventional Forces Europe (CFE) Negotiations*

The Soviet proposal was remarkably similar in its approach to initial reductions, but was more ambitious in its scope and long-term objectives. Soviet negotiators opened with a three stage proposal: (1) a two-three year period during which both sides would reduce offensive weapons to levels 10–15 per cent below the lowest level possessed by either side. The largest reductions were proposed for the two Germanies where there would be a total ban on nuclear weapons; (2) a second three-year phase would reduce arms by an additional 25 per cent; (3) the final stage, lasting to the year 2000, would have both alliances restructuring their forces for 'purely' defensive capabilities.[33]

The Soviet proposal is significant for both arms control and NATO conventional modernization strategy. The devil and years of negotiations are in the details, but a final arms control and verification regime must not only reduce instability at the central front, but also in the reinforcing zone where forces could be deployed for rapid reinforcement of a surprise attack. Force levels in one zone may be determined to some degree by the ultimate disposition of men and weapons that are removed from another. Will, for example, Soviet troops and divisions be removed from the force structure? Will their weapons and equipment be stored West of the Urals, East of the Urals, or destroyed? The vague outline of non-offensive defense has not addressed these specific problems. Anticipating lengthy negotiations on these and other questions, NATO conventional force modernization should proceed. Many decisions can be made and a considerable degree of modernization can precede a conventional arms treaty.

Options described here are not intended to be taken as narrow prescriptions or as criticisms of either side's proposals at the negotiations. There are many possible variants to general principles. One approach summarized in Table 5.5 is to link modernization strategy to the arms control zones depicted in Map 5.1. Modernization in the central front should support conventional strategy and develop maximum firepower and mobility per unit of manpower. A credible conventional deterrence and alternative defensive concepts are needed to exploit Soviet force reductions through maximum deployment of wide-area, high-tech submunitions deliverable from the new Multiple Launched Rocket System (MLRS), aircraft, and the Army's short-range tactical missile system (ATACMs).

These systems, together with other forces deployed during the Reagan buildup (M-1 Tanks, Bradley Fighting Vehicles, Black Hawk

TABLE 5.5 *Strategy After Conventional Arms Control*

Zone	Modernization strategy	Deterrence strategy
• Central front	Conventional forces/'smart' munitions with wide area coverage	Conventional deterrence
• Deep central front	Maintain existing theater nuclear force; long-term R & D for long-range conventional forces	Flexible response without requirement for nuclear first-use
• Reinforcing zone		Strategic nuclear deterrence
• Reserve zone	Strategic nuclear modernization with more emphasis on survivable weapons; less emphasis on hard target counter-force capabilities	Flexible options to attack Soviet conventional forces in the reinforcing zone
• Continental US		Deter Soviet use of strategic and theater nuclear weapons

Helicopters, and Patriot Air Defense Missiles), more than double the firepower of every American division. 'Brilliant' munitions in development and emerging technologies (lasers and kinetic energy weapons) promise, as Marshal Ogarkov predicted, to give conventional forces on the defensive the same degree of lethality as battlefield nuclear weapons.

Theater nuclear weapons have been required to threaten critical targets deep in Eastern Europe. Airfields and the rail transshipment points along the Soviet-East European border are especially vital to sustain Soviet military momentum. Rail transshipment points are the bottlenecks created by Soviet construction of tracks that are wider than their European counterparts, an anomaly that requires off-loading Soviet trains and reloading cargoes on European trains. Broad-gauged rails have been described as a strategic measure to hinder an invasion of Russia. In fact, the original recommendation was made by an American technical adviser to the Czar as the most cost-effective means to support high volume rolling stock and to ensure stability at high speed. Ironically, the Russian Civil War and World War II demonstrated that while variations in rail gauges did slow the logistical support of more rapidly advancing military forces, it was easier for invaders to re-lay one rail on Russia's broad gauge track than for the Russians to widen European tracks.[34] These self-imposed bottlenecks and the long, fixed rail routes through the Soviet Union make their reinforcement of Europe no less arduous than Western defenses of the sea-lanes, ports and NATO airfields. A long-range research and development program should be pursued to put these choke-points at risk with conventional weapons. Early use of nuclear weapons on or near the Soviet border in support of Airland Battle is a potential escalator that may result in political indecisiveness and dangerous delays in striking critical targets. Under such conditions, conventional deterrence is more credible than nuclear deterrence.

The most divisive decision confronting NATO is the modernization of short-range nuclear forces for the European battlefield. The last ground-to-ground nuclear missile, the Lance, will be phased out in the 1990s. The administration seeks approval of a program to modernize its arsenal of short-range nuclear weapons. The options include: (1) development of a new missile with a reported range just under the ceiling established by the INF Treaty which bans ground-launched missiles with a range of 300 to 3400 miles; (2) a new air-launched missile similar to the SRAM, an air-to-surface missile carried by strategic nuclear bombers; and (3) continued production of modernized nuclear artillery shells.[35]

It is unlikely that either Congress or the NATO allies will support full development or deployment of these systems. There is a strong political opposition in West Germany along with growing support for triple zero–the elimination of all remaining nuclear weapons on the central front. German political rhetoric is illustrative of the problem: the shorter the range of the weapon, the deader the Germans.

Conventional arms control negotiations could be seriously disrupted by a divisive debate within NATO over nuclear modernization. The issue puts the horse before the cart in the sense that the general outcome of an arms control treaty and conventional force modernization should precede a final decision on new nuclear weapons. Reductions of Soviet armored and mechanized divisions and NATO conventional modernization may serve the same strategy that theater nuclear forces once served. That is to put at risk any Warsaw Pact forces that mass for an attack along the central front. If conventional modernization produces weapons capable of lethality over the breadth and depth of the battlefield in support of NATO's forward defense and the Airland Battle doctrine described in Chapter 1, the case for nuclear modernization is significantly weakened. The primary deficiency in current programs is the short-range of conventional weapons. They have the lethality to disrupt a Soviet attack, but they lack the range to fully supplement air strikes against Soviet second echelons. Arms control may succeed in reducing these threats and political will can produce long-range, lethal conventional weapons. Current munitions for the Army's Multiple Launch Rocket System (MLRS) have a range of 45 kilometers. The new ATACMs will extend that to well over 100 kilometers, coinciding with the 50–150 kilometers prescribed by the Airland Battle doctrine to engage Soviet second echelons. The trade-off between conventional and nuclear modernization should be weighed against both the military requirements for disrupting Soviet momentum on the battlefield and the political requirements for NATO's cohesion. It is by no means clear that nuclear modernization is the best means for accomplishing either objective.

A warning by former German Chancellor Helmut Schmidt is instructive. Schmidt wrote that he had confidence in conventional defenses, even though:

> the strategy of flexible response has always implied a quick escalation toward very early first use of nuclear weapons by the West. But it is unrealistic to believe that West German soldiers would fight after the explosions of the first couple of nuclear weapons on West German soil; the West Germans would certainly not act anymore fanatically or suicidally than the Japanese did in 1945 after Hiroshima and Nagasaki.[36]

It is difficult to find a more eloquent argument for conventional deterrence in Central Europe. Schmidt concluded that nuclear weapons are valuable only to deter Soviet nuclear use, not as instruments to deter limited war or even large-scale conventional attack.

Anti-nuclear sentiment reached a peak under Chancellor Helmut Kohl and Foreign Minister Hans-Dietrich Genscher. Domestic politics in the Federal Republic make it impossible to modernize short-range nuclear forces (SNF) outside formal Soviet-American negotiations to limit their numbers in the European theater. Veteran arms control negotiator Paul Nitze endorsed the German position and formal negotiations on SNF to achieve a balance in a category of weapons in which the Soviets are dominant and to avoid exacerbating political divisions in Germany at a time of growing impatience with the extraordinary concentration of foreign armies and weapons on their soil.[37]

Political pressure from the Germans resulted in a compromise similar to the Nitze proposal. President Bush's broad arms control offensive at the 40th NATO anniversary summit in Brussels opened the door to a compromise solution to the SNF issue.[38] In their joint communiqué, NATO heads of state reaffirmed their commitment to a 'strategy of deterrence based upon an appropriate mix of adequate and effective nuclear and conventional forces'. At the same time, the allies understand that 'negotiated reductions leading to a level below the existing level of their SNF missiles will not be carried out until the results of these negotiations [CFE] have been implemented'.[39] The compromised language rules out for the immediate future the 'triple zero' option preferred by the Germans.[40]

The problem that hangs over the negotiations, however, is the extent to which the Soviets will continue to press the Germans on the issue of SNF. The Soviets' German strrategy is tied to the broader objective of a denuclearized Europe (that is, all land-based systems, including dual-capable aircraft). The strategy exploits German fears of 'singularization'. 'Singularization' is the German geographic predicament of being the battlefield for a majority of nuclear weapons that were not elimintated by the INF Treaty (Lance missiles and artillery-fired atomic projectiles). The fact that these weapons are 'German killers' (in the geographic sense) is a source of great discomfort to our most important NATO ally, and no doubt a source of some cynical pleasure in the minds of Soviet strategists.

During a visit to Bonn soon after the signing of the INF Treaty, Soviet Foreign Minister Edward Shevardnadze pressed his German hosts for their support of the Soviet 'triple zero' proposal. Triple zero not only fans the German fears of singularization, but could also result in a more credible Soviet conventional war option in Europe.[41] For that reason, neither the Reagan nor the Bush administrations have been willing to accept the triple zero option prior to firm Soviet agreements on conventional reductions. Over the long term, however, German domestic politics will almost certainly require triple zero for land-based nuclear missiles and artillery.

The remaining geographic zones identified in Map 5.1 can be defended with discriminate strategies and forces. NATO's northern and southern flanks (primarily Norway and Turkey) should, like the central front, depend on conventional deterrence that is decoupled from threats to set off a rapid chain of nuclear escalation. Strategy in the reinforcing zones–Great Britain, France and Italy on the one side and the Soviet military districts adjacent to Eastern Europe on the other, should remain independent of conventional deterrence in the central zone. Deterrence in the reinforcing zones should rest unambiguously on the theater and strategic nuclear forces and strategy described previously in this chapter. Escalation of war to these zones risks full-scale theater strategic war and should be deterred by the same levels of threat used to deter intercontinental nuclear war.

Linking arms control and conventional and strategic nuclear modernization to specific zones in the Atlantic to the Urals region does not mean that the US commitment to extended deterrence varies from one ally to another. The distinctions mean that conventional deterrence is possible far below the nuclear threshold.

Ironically, if negotiations produce a treaty, US conventional forces will become strategically more important. If their redeployment to the United States results in demobilization, conventional deterrence will be weakened. Total manpower may decline, but the number of army divisions (18 active and 10 reserve) should be retained, even expanded through organizational devices similar to the Soviet practice of maintaining ground forces at various readiness levels. These categories range from full-strength combat ready divisions, to cadre divisions with less than 50 per cent of their required manpower, to small custodial forces for weapons and equipment. National mobilization and training are required to bring them to full strength, but even on paper they broadcast to the world a level of commitment and a corresponding component in the structure of deterrence.

Land power is unique in the level of national will and commitment that it reflects. Naval and air power are essential components in US defense posture and conventional deterrence, but they are also the symbols of limited commitment. They are more prone to substitute service strategies–air power or maritime strategy–for national strategy. Land power is more closely identified with and dependent on national strategy because it is the symbol of the nation's highest commitment of military power short of nuclear weapons.[42] When the nation commits its army, the commitment is nearly always total, and the cost of failure far more damaging to national prestige.

The Soviet problem is more economic than military. Soviet military forces must be reduced to finance economic reform. Skeptics in the

West should not underestimate the risks this entails for Gorbachev. The Soviet Union depends disproportionately on its military might for superpower status. Previous Soviet leaders have assumed the convertibility of military assets to diplomatic, economic, and psychological gains consistent with Soviet desires to extend their influence. The size and sophistication of Soviet forces are the most visible product of industrial modernization. They convey the symptoms of success. In Soviet eyes, respect and authority must certainly spill over to their political and ideological claims. Gorbachev is openly challenging these sacred assumptions. Security, he has argued, and by inference superpower status, cannot rest on military power alone. Political and economic co-operation with the West is an essential part of State security in the nuclear age. His recognition of the limited utility of military power is a giant step toward a post-nuclear world.

Yet Europe remains the most militarized zone in the world. There is growing fear among Europeans that preparations for war and the infrastructure of deterrence itself have become the greater threat. Mutual disengagement with disproportionate reductions on the Soviet side can reduce the political tensions that have persisted since two powerful allies met on the Elbe in 1945. The continued presence of American and Soviet armies in Central Europe for more than 45 years after World War II is neither inevitable nor a natural part of international politics. Powerful allies in Western Europe and Gorbachev's revolutionary attempts to reform the Soviet Union are dramatic symbols of success for American post-war strategy in Europe. The challenge in the next century is learning how to live with that success.

NOTES

1. Compare, for example, the 1976 and the 1986 editions of FM 100–5, *Operations*, the Army's manual for disseminating official doctrine.
2. John Scott, *Terminating Major War in Europe*, (Carlisle, Pa.: Strategic Studies Institute, US Army War College, 1989).
3. Deputy Under-Secretary of Defense Lawrence W. Woodruff, 'Nuclear Force Modernization Efforts', *Defense Issues*, 1 March 1988, pp. 1–8.
4. Congressman Les Aspin, Chairman, House Armed Services Committee, *Washington Post*, 12 January 1989, p. A26.
5. Richard Wolkomie, 'The Once and Future Bomber', *Air and Space*, Vol. 3, No. 6, February/March 1989, pp. 70–3.
6. The original specification for the B-1B bomber provided a 22 ALCM capacity. However, flight control problems have forced the Air Force to reduce loads because external cruise missile carriers create unex-

pected drag that reduces the bomber's range. The defensive avionics systems required for the bomber's penetration mission remains the most serious deficiency. Nearly all the aircraft's technical problems are related to its penetration mission. See *Washington Post*, 4 February 1989, p. A1.

7. Statement by former Chief of the General Staff, Marshal Akhromeyev, at the Reykjavik Summit. Quoted in Strobe Talbott's, 'Why START Stopped', *Foreign Affairs*, Vol. 67, No. 1, Fall 1988, p. 61.
8. Freeman Dyson, *Weapons and Hope* (New York: Harper & Row, 1984), Chapters 5 and 7.
9. Data is from *The Military Balance, 1988–1989* (London: International Institute for Strategic Studies, 1988), pp. 30, 33, 36, 216.
10. The CIA has estimated that the Soviets may deploy from 2000–3000 nuclear cruise missiles by the mid-1990s. This estimate included ground-launched cruise missiles now banned by the INF Treaty. Figures quoted in Peter Clausen, *et al.*, *In Search of Stability: An Assessment of New U.S. Nuclear Forces* (Union of Concerned Scientists: Washington, DC, 1986), p. 62.
11. Mikhail Gorbachev, 'Outer Space Should Serve Peace', reprinted from *Pravda*, 6 July 1985 in *Information Bulletin*, Vol. 23, No. 18, 1985, p. 4; and *Star Wars: Delusions and Dangers* (Moscow: Military Publishing House, 1985), p. 10.
12. "Geneva: What Has the First Rounds of Talks Shown?" *Information Bulletin*, Vol. 23, No. 14, 1985, p. 46.
13. Gary L. Guertner and Donald Snow, *The Last Frontier: An Analysis of the Strategic Defense Initiative* (Lexington, Ma.: Lexington Books, 1986), Chapter 1.
14. Examination of Soviet Space and Ballistic Missile Defense programs can be found in *Soviet Military Power*, 1985–88 edns (Washington, DC: US Department of Defense); Nicholas L. Johnson, *Soviet Military Strategy in Space* (London: Jane's Publishing, 1987); and Stephen M. Meyer, 'Space and Soviet Military Planning', in William J. Durch (ed.), *National Interest and the Use of Space* (Cambridge, Mass.: Ballinger, 1984), Chapter 3, pp. 61–88.
15. Many excellent histories of this evolution have been published recently. This section draws upon David Rosenberg, 'The Origins of Overkill: Nuclear Weapons and American Strategy, 1945–1960', *International Security*, Vol. 7., No. 4, Spring 1983,pp. 3–71; Fred Kaplan, *The Wizards of Armageddon* (New York: Simon & Schuster, 1983); and Desmond Ball, *Targeting for Strategic Deterrence*, Adelphi Paper 185 (London: International Institute for Strategic Studies, 1983).
16. William F. Burrows, *Deep Black: Space Espionage and National Security* (New York): Random House, 1986), Chapters 4, 6, and 9.
17. Richard Nixon, *U.S. Foreign Policy for the 1970s: A New Strategy for Peace*, a report to Congress, 18 February 1970 (Washington, DC: US Government Printing Office), p. 122.
18. James Schlesinger, Address for Writer's Association Luncheon, DOD Public Affairs Office, 10 January 1974, pp. 5–6.

19. Desmond Ball, *Targeting for Strategic Deterrence*, Adelphi Paper (London: International Institute for Strategic Studies, 1983), p. 23.
20. A summary of this evolution is described in Leon Sloss and Marc Dean Millot, 'US Nuclear Strategy in Evolution', *Strategic Review*, Vol. XII, Winter 1984, pp. 19–25.
21. *Washington Post*, 20 March 1987, p. A17.
22. Scott D. Sagan, 'SIOP-62: The Nuclear War Plan Briefing to President Kennedy', *International Security*, Vol. 12, No. 1, Summer 1987, pp. 22–51.
23. Frank von Hippel, *et al.*, 'Civilian Casualties From Counterforce Attacks', *Scientific American*, Vol. 259, No. 3 (September 1988), pp. 36–42. In 1974 Secretary of Defense James Schlesinger told the Senate Foreign Relations Committee that a 'limited' Soviet attack on the United States against 2158 strategic nuclear targets would produce an estimated 6.7 million dead. A 1975 Department of Defense study increased that estimate to 21.7 million dead. What is common to all studies is the large number of nuclear weapons required to conduct a 'limited' counterforce attack. These studies are described in the *Los Angeles Times*, 17 September 1975, p. 9.
24. Von Hippel, op. cit., p. 42.
25. Philip A. Karber, 'Conventional Arms Control, or Why Nunn is Better Than None', in Uwe Nerlich and James A. Thompson (eds), *Conventional Arms Control and the Security of Europe* (Boulder, Co.: Westview Press, 1988), p. 174.
26. Delegates from the 35 members of the Conference on Confidence and Security-Building Measures and Disarmament in Europe (CDE) agreed on several on-site inspection provisions in 1986. NATO and Warsaw Pact military commanders may not train or exercise units in excess of 300 tanks or 13 000 personnel without prior notification. When exercises exceed 17 000 personnel, military representatives from the other side must be allowed on-site observation of the exercise. Under CDE provisions, American and British personnel have observed Warsaw Pact exercises in the Soviet Union, Hungary, and the German Democratic Republic. Soviet personnel have been present during NATO exercises in West Germany, Turkey, Norway, and Great Britain.
27. Michael Moody, 'Conventional Arms Control: An Analytical Survey of Recent Literature', *The Washington Quarterly*, Vol. 12, No. 1, Winter 1988.
28. *New York Times*, 31 January 1989, p. 1.
29. *Combat History Analysis Study Effort (CHASE)*, Progress Report, August 1986, Requirements and Resources Directorate, US Army Concepts Analysis Agency, pp. 3–20.
30. Philip A. Karber, 'The Military Impact of the Gorbachev Reductions', *Armed Forces Journal International*, January 1989, pp. 54–64.
31. Sam Nunn, 'NATO Challenges and Opportunities: A Three-Track Approach', *NATO Review*, June 1987, pp. 1–7.
32. The Soviet zones are described by Oleg Amirov, *et al.*, 'Conventional

War: Strategic Concepts', in *Disarmament and Security 1987 Yearbook* (USSR Academy of Sciences: Moscow, 1988), pp. 395–400. The NATO Plan is described in *Army Times*, 8 May 1989, p. 16. Map 1 is derived from both proposals, which vary slightly in their zonal boundaries.

33. The initial proposals by NATO and the Warsaw Pact at the Conventional Forces in Europe Negotiations (CFE) are summarized in *Defense News*, 13 March 1989, p. 1.
34. J. N. Westwood, *A History of Russian Railways* (London: Allen & Unwin, 1964), pp. 29–31 and 237–41. Soviet rails are gauged 5 feet apart. Their European counterparts are 4 feet, eight and one-half inches. The $3\frac{1}{2}$ inches have great strategic significance.
35. *Washington Post*, 17 February 1989, p. 1.
36. Department of Defense, *Current News*, 15 July 1987, p. 11. Reprinted from *Fort Worth Telegram and Star*, 3 May 1987.
37. Paul Nitze, Press Conference on his final day of service in the Bush Administration, *New York Times*, 8 May 1989, p. 1.
38. The President reversed American opposition to cutting aircraft and troop strength. He offered to bring 30 000 US troops home, leaving 275 000 in Europe if the Soviets would reduce their European forces to an equal number. Troop cuts would be accompanied by a 15 per cent cut below current NATO levels in all types of aircraft. *New York Times*, 30 May 1989, p. 1.
39. *New York Times*, 31 May 1989, p. A15.
40. A public opinion poll taken prior to the NATO summit found that 89.1 per cent of those Germans questioned opposed American short-range nuclear weapons. *New York Times*, 27 May 1989, p. 6.
41. German fears have been voiced on many occasions by Chancellor Helmut Kohl and members of his government. One advisor has suggested reducing warheads for battlefield nuclear weapons by 50 per cent, with the majority of cuts coming from *short-range* nuclear artillery. The emphasis on short-range weapons illustrates German sensitivities on where the battlefield will be. See *Washington Post*, 29 February 1988, p. A6 and 27 February 1988, p. A13. For this reason, the Germans are unlikely to support Soviet calls for eliminating nuclear weapons carried on tactical aircraft. See the joint Soviet-Czechoslovak communiqué, printed in *FBIS*, 15 April 1987, p. F20. It is also worth noting that 'non-offensive defense' is rooted in the German peace movement. For examples of literature from the German Peace Movement see *The Journal of Peace Research*, Vol. 24, No. 1, March 1987.
42. The author is grateful to Carl H. Builder for these interservice comparisons. See his *The Army in the Strategic Planing Process: Who Shall Bell the Cat?* (Santa Monica,Ca.: RAND, 1987), Chapters 3, 4, and 6.

Index

ABM Treaty, 97–9, 108–11, 113, 115–17, 122, 124, 133, 144
ACMs (advanced conventional munitions), 58
AirLand Battle, 21, 168
Akhromeyev, General Sergei F., 68–9
Andropov, Yuri, 60, 85–6
ASAT, 124, 139
atomic monopoly, 39, 148–9

bombers, strategic
  B–1B, 138–9, 142, 158
  B–2 (Stealth), 138, 142–3, 158
  B–52, 42, 138–9, 142
  Bear, H. 118, 146
  Blackjack, 146, 158
Bradley, General Omar, 7, 14
Brezhnev, Leonid, 35–6, 38–56, 85, 100, 103
Brown, Harold, 124

Carlucci, Frank C., *Fn* 6, 138
Carter, Ashton, 154
center of gravity,
  Soviet, 92, 159, 162
  NATO, 159, 161
Chernenko, Konstantin, 60, 115
coalition deterrence, 9, 24
Conventional Forces in Europe Negotiations (CFE), 67–9, 159–71, data base for, 161–2, zonal approach to, 163–4, 166, 169–70
competitive strategy, 138
  see also Soviet vulnerability
countervalue targets, 154, 156
counterforce targets, 154, 156
Cuban missile crisis, 50

D–5 missile, 137, 146, 158
deterrence
  nuclear, 2–4, 15, 19, 25
  conventional, 15–16, 18, 25
deterrence by denial, 45, 156
deterrence by punishment, 2, 156
Dulles, John Foster, 42, 150

ethnic vulnerability (of Soviet Union) 85, 94
Eisenhower, Dwight, 149–50
extended deterrence, 13, 29, 42, 55
  see also strategic coupling

flexible response, 3, 7–10, 13–29
Follow-on Forces Attack (FOFA), 19–22
forward defense, 10–14

Gerasimov, Gennady, 65
Glasnost, 61, 86, 88
Gorbachev, Mikhail, 34, 60, 62–71, 79, 86–8, 145, 162–3, 170–1
Gray, Colin, 92
Grechko, Marshal, 55, 59

homeland defense (Soviet), 79–80

INF Treaty, 4, 28, 62–4, 167

Karber, Philip, 160
Kissinger, Henry, 54
Khrushchev, Nikita, 41, 43–51, 69, 85, 103
Krasnoyarsk radar, 101, 108–9, 115–17, 122–4

Lambeth, Benjamin, 38, 91
Lemnitzer, General Lyman L., 155

limited nuclear war, 3, 38, 93, 150–7

McNamara, Robert, 9, 14–17, 65–6, 150–4
Meyer, Stephen, 35, 104
Multiple Launch Rocket System (MLRS), 167–8, 142, 165
mutual deterrence, 135–7, 154–8
  see also strategic stability
MX missile, 137, 140–1, 146, 156, 158

NATO strategy, 2–4, 7–29, 133–4, 165–7
Nitze, Paul, 16, 168
Nixon, Richard, 152
nuclear targeting, 148–56
Nunn, Senator Sam, 28, 163

Ogarkov, Marshal N. V., 55–6, 58–60
on-site inspection, 127, 160
Operational Maneuver Groups (OMGs), 21–2, 56–7

parity, Soviet achievement of, 50–3
Perestroika, 61, 66
Pershing II ballistic missile, 25–6, 29, 62–3
Perle, Richard, 126
post-nuclear world, definition of, 1

Quelle, Henri, 13

rail transshipment points, 26, 166–7
Reagan, Ronald, 1, 97–100, 125–6, 133–7, 147, 152–3, 165–6
reasonable sufficiency, 64–71
Rogers, Bernard, General, 18, 26, 28

SS-9 missile, 52–3
SS-11 missile, 52–3, 118, 145
SS-13 missile, 52–3
SS-16 missile, 107, 127
SS-18 missile, 53, 65, 123, 140–1, 145–6, 156, 158
SS-20 missile, 61–3, 119

SS-24 missile, 53, 65, 114, 145, 158
SS-25 missile, 53, 65, 101, 107, 110, 114–15, 119–23, 145, 158
SALT I Treaty, 98, 100–1, 107–11, 120, 122
SALT II Treaty, 97, 125
Schlesinger, James, 151–2
Schmidt, Helmut, 168
Scott, John, 137
Scowcroft Commission, 144
sea-launched cruise missiles, 138–9, 143
Sokolovsky, Marshall Sergei, 43, 49, 54
Single Integrated Operation Plan (SIOP), 17, 150
Soviet arms control bureaucracy, 100–6, 125
Soviet military doctrine, 37–8, 47, 49, 51
  see also homeland defense
Soviet non-compliance (with arms control agreements), 97–129
  conflicting interpretation of, 99–120
  military significance of, 99, 121–4
Short-range Nuclear Missiles (SNF), 167–9
  see also Triple ZERO
Soviet military strategy, 21, 26, 34, 37–8, 61, 68
Soviet nuclear artillery, 57
Soviet views of deterrence, 26, 37–71
Soviet view of flexible response, 8–9, 51, 55
Soviet vulnerability, 80, 85–95
  see also competitive strategy
Stalin, Joseph, 39–45, 83
Strategic Air Command (SAC), 46, 149–50
Strategic Defense Initiative (SDI), 133, 135–7, 144–5, 147–8
Standing Consultative Committee (SCC), 113, 118, 125
START Treaty, 65, 135–6, 139, 140–7
strategic coupling, 3, 17, 25–7, 29, 42, 55

Index

Strategic Rocket Forces (SRF), 46
strategic stability 134–40, 144–8, 154–8
see also mutual deterrence

Talensky, General A., 43
Tokaev, Colonel G. A., 41
Trident submarine, 137, 141
Triple ZERO, 169
see also short-range nuclear missiles

Truman, Harry, 41, 149
Tula speech, 35–6, 38, 65

Ustinov, Dimitry, 59–60
Weinberger, Casper, 97, 121, 138
Wolfe, Thomas W., 50

Yazov, Demitri, 68